Current Topics in Microbiology 190 and Immunology

Editors

Borna Disease

Edited by H. Koprowski and W. I. Lipkin

With 33 Figures and 6 Tables

Springer-Verlag Berlin Heidelberg GmbH

Hilary Koprowski
Department of Microbiology and Immunology
Head of Center of Neurovirology
at Thomas Jefferson University
Philadelphia, PA 10107, USA

W. Ian Lipkin
Department of Neurology
College of Medicine
University of California, Irvine
Irvine, CA 92717–4290, USA

Cover Illustration: Borna disease virus originally discussed as an infection of horses and sheep is now recognized as affecting many warmblooded animals, including ostriches. Possible infection of man cannot be ruled out.

Cover design: Künkel und Lopka, Ilvesheim

ISBN 978-3-642-78620-4 ISBN 978-3-642-78618-1 (eBook)
DOI 10.1007/978-3-642-78618-1

Originally published by Springer-Verlag Berlin Heidelberg New York in 1995
Softcover reprint of the hardcover 1st edition 1995

Production: PRODUserv Springer Produktions-Gesellschaft, Berlin
Typesetting: Thomson Press (India) Ltd, New Delhi
SPIN: 100 96 427 27/3020-5 4 3 2 1 0 – Printed on acid-free paper.

Preface

Borna disease (BD) was first described over 200 years ago, in what is now south-eastern Germany, as a fatal neurologic affliction of horses. A veterinary handbook published in 1785 contained an accurate report of the clinical syndrome, implicated unsatisfied sexual desires or overfeeding in its pathogenesis and suggested such colorful therapies as phlebotomy, plucking hair from selected locations, and threading a rope coated with Spanish fly cream through the subcutaneous tissues of the infected horse [1]. Though the pathogenesis of BD is now better understood, we still have no effective methods for preventing or treating the illness.

The importance of BD has increased with the recognition that the geographic distribution and host range are larger than previously appreciated. In addition to horses, natural disease has been reported in sheep, cats and ostriches. Antibodies reacting with BDV antigen(s) were recorded in animals in Europe, North America, Asia and Africa. Though infectious virus has not been isolated from human subjects, several groups in North America and Europe have found antibodies reactive with BD virus (BDV) in patients with neurologic and psychiatric disorders. As described in this volume, BDV proteins have been found in blood monocytes obtained from these patients. It is thus possible that humans suffering from psychiatric disorders were either infected with BDV or react immunologically to an antigen closely related to BDV.

The field of BD research continues to evolve rapidly. We have tried in this volume to choose contributions which emphasize aspects of the field that should be of enduring interest to the reader. Authors have been encouraged to present data in press to enhance the probability that information will be current at the time of publication. Nevertheless, because of the rapid progress in the field of BDV research it has been particularly difficult to keep abreast of advances in the molecular biology of BDV. Thus, by the time this volume is published, more information may have become available concerning the nature of the BDV genome, and replication and transcription of the virus. Fortunately, these advances will complement rather than contradict the information already presented.

We are grateful to the authors for their efforts. W.I.L. wishes to thank Stanley and Evelyn Lipkin and Michael Oldstone for encouragement and support.

HILARY KOPROWSKI
W. IAN LIPKIN

Reference

Abildgaard PC (1785). Pferde- und Vieharzt in einem kleinen Auszuge; oder, Handbuch von den gewöhnlichsten Krankheiten der Pferde, des Hornivichee, der Schafe und Schweine, sammt der bequemsten und wohlfellesten Art sie zu heilen. Zum Gebrauch des Landmanns. (Eds. Johann Thomas Edlen von Tratthern, Wien)

List of Contents

T. Briese, W.I. Lipkin, and J.C. de la Torre:
Molecular Biology of Borna Disease Virus 1

R. Rott and H. Becht:
Natural and Experimental Borna Disease in Animals . . . 17

M. Malkinson, Y. Weisman, S. Perl, and E. Ashash:
A Borna-Like Disease of Ostriches in Israel. 31

G. Gosztonyi and H. Ludwig:
Borna Disease—Neuropathology and Pathogenesis . . . 39

L. Stitz, B. Dietzschold, and K.M. Carbone:
Immunopathogenesis of Borna Disease 75

M.V. Solbrig, J.H. Fallon, and W.I. Lipkin:
Behavioral Disturbances and Pharmacology
of Borna Disease 93

L. Bode:
Human Infections with Borna Disease Virus
and Potential Pathogenic Implications. 103

Subject Index 131

List of Contributors

(Their addresses can be found at the beginning of their respective chapters.)

Ashash E. 31

Becht H. 17
Bode L. 103
Briese T. 1

Carbone K.M. 75

Dietzschold B. 75
de la Torre J.C. ... 1

Fallon J.H. 93

Gosztonyi G. 39

Lipkin W.I. 1, 93
Ludwig H. 39

Malkinson M. 31

Perl S. 31

Rott R. 17

Solbrig M.V. 93
Stitz L. 75

Weisman Y. 31

Molecular Biology of Borna Disease Virus

T. Briese[1,2], W.I. Lipkin[1,3], and J.C. de la Torre[4]

1 Introduction 1

2 Genetics 2
2.1 Characterization of BDV Nucleic Acids 2
2.2 BDV Contains Genomic RNA of Negative Polarity 5

3 Transcription 6
3.1 Transcription of BDV Occurs in the Nucleus of the Infected Cell 6

4 Proteins 8
4.1 Borna Disease-Specific Proteins 8
4.2 Analysis of Cloned Viral Proteins 9

5 Concluding Remarks 11

References 13

1 Introduction

Originally described in the early nineteenth century as a fatal encephalitis in horses, Borna disease (BD) has become an extraordinarily valuable model for the study of both molecular mechanisms and biological consequences of persistent virus infection in the CNS (Nicolau and Galloway 1928; Zwick 1939, this volume). BD is an immune-mediated neurologic syndrome characterized by behavioral abnormalities, meningeal and parenchymal inflammatory cell infiltrates in the brain, and the accumulation of disease-specific antigen in limbic system neurons (Joest and Degen 1911; Seifried and Spatz 1930; Ludwig et al. 1988; Richt et al. 1992). As a natural infection, BD has only been confirmed to occur in horses and sheep, but experimentally it can be transmitted to an extraordinary wide range of host species, including birds, rodents and nonhuman primates (Zwick et al. 1926; Nicolau and Galloway 1928; Zwick 1939; Matthias 1954; Nitzschke 1963; Heinig 1969; Anzil et al. 1973; Ludwig et al. 1973; Metzler et al. 1976; Ludwig and Thein

[1] Department of Neurology, University of California, Irvine, CA 92717, USA
[2] Institute of Virology, Freie Universität Berlin, Nordufer 20, 13353 Berlin, Germany
[3] Department of Anatomy and Neurobiology, University of California, Irvine, CA 92717, USA
[4] Department of Neuropharmacology, Scripps Research Institute, 10666 North Torrey Pines Road, La Jolla, CA 92037, USA

1977; SPRANKEL et al. 1978; STITZ et al. 1980; HIRANO et al. 1983, NARAYAN et al. 1983; GOSZTONYI and LUDWIG 1984; KAO et al. 1984; LUDWIG et al. 1985; WAELCHLI et al. 1985; RICHT et al. 1992). Serological data indicate that the host range may even extend to humans, although no infectious material has been isolated from human subjects (AMSTERDAM et al. 1985; ROTT et al. 1985, 1991; BODE et al. 1988, 1992).

Despite remarkable progress achieved during the past decade in understanding the pathogenesis of BD, until quite recently little was known about the etiologic agent for this disease (LUDWIG et al. 1988; DUCHALA et al. 1989). Biochemical studies indicated that BD is induced by an agent which is sensitive to detergents, organic solvents and UV light (NICOLAU and GALLOWAY 1928; DANNER and MAYR 1979; DUCHALA et al. 1989). In addition, a size of 80–125 nm for the agent was estimated, based on filtration experiments (ELFORD and GALLOWAY 1933; DANNER and MAYR 1979). These results indicated that BD is likely to be caused by a conventional, enveloped virus, the Borna disease virus (BDV). However, such features as noncytopathic replication, low titers of infectious material and tight cell association hampered isolation and further characterization of this agent (LUDWIG et al. 1988). The detection of BDV relied primarily upon indirect, immunologic methods (WAGNER et al. 1968; GOSZTONYI and LUDWIG 1984; PAULI et al. 1984). BD induces the production of specific proteins, the "s-antigen," which elicit a strong immune response (VON SPROCKHOFF 1956; LUDWIG and BECHT 1977). Polyclonal serum antibodies and oligoclonal antibodies in the CNS directed against s-antigen components of 38/40 kDa and 24 kDa have been characterized (LUDWIG et al. 1977, 1988). Immunohistochemical analysis indicated that the agent is highly neurotropic and spreads intra-axonally (KREY et al. 1979; CARBONE et al. 1987; MORALES et al. 1988). Definitive proof of the viral nature of the agent came only after application of molecular genetic approaches, which led to the isolation of BDV-specific cDNAs and a partial characterization of this virus (LIPKIN et al. 1990; VANDEWOUDE et al. 1990).

In this chapter, we summarize the present knowledge of the molecular biology of Borna disease virus, the causative agent of Borna disease.

2 Genetics

2.1 Characterization of BDV Nucleic Acids

The isolation of BDV-specific cDNAs by using subtractive cloning procedures has provided both direct evidence for an infectious basis for BD and a battery of tools for molecular characterization of BDV. Two classes of cDNAs corresponding to each of the two major antigens known to be specific for BD were isolated from BDV-infected rat brain: one class corresponding to a 40 kDa protein, open reading frame (ORF) p40, the other to a 24 kDa protein, ORF p24 (LIPKIN et al. 1990). cDNAs of the latter class were also isolated from BD-infected tissue culture cells through subtractive cloning or through oligonucleotide screening using sequences deduced from direct amino acid sequencing of purified BD antigen (VANDE WOUDE et al. 1990; THIERER et al. 1992).

Northern blot experiments using these cDNA probes revealed RNAs of 0.8–1.2 kb, 2.1 kb, 3.5 kb and approximately 8.5 kb, exclusively present in extracts obtained from BDV-infected material (Fig. 1) (LIPKIN et al. 1990; DE LA TORRE et al. 1990; VANDEWOUDE et al. 1990). In Southern hybridization studies, the cDNA probes did not hybridize to genomic DNA, ruling out the possibility that the cloned sequences were of cellular origin (LIPKIN et al. 1990; VANDEWOUDE et al. 1990). Accordingly, the pattern of BDV nucleic acid signal found in in situ hybridization studies of BD rat brain was not evenly distributed over all cells, but was instead consistent with the well characterized distribution of BD antigen in infected rat brain (see also Gosztonyi and Ludwig, this volume). Hybridization signal was found in layer 4 and 5 cortical neurons and in brainstem, with higher density over limbic structures including thalamus and sectors CA3, CA4 of hippocampus (LIPKIN et al. 1990; CARBONE et al. 1991; RICHT et al. 1991). Because of this hetero-geneous distribution it was necessary to address the question whether BDV is a DNA or RNA virus, or a retrovirus in homogeneously infected cell populations by using persistently BDV-infected cell lines. Immunofluorescence as well as in situ hybridization analysis showed that in the MBV line, a persistently BDV-infected cell line derived from MDCK cells, all the cells were infected (DE LA TORRE et al. 1990). BDV sequences were not detected in genomic or episomal DNA obtained from these cells, using conditions that allowed the detection of 0.5 gene copies per cell genome equivalent (Fig. 1) (DE LA TORRE et al. 1990). These results demonstrated that BDV is neither a DNA virus nor a retrovirus (DE LA TORRE et al. 1990; VANDE WOUDE et al. 1990). The sensitivity of the BDV-specific nucleic acids to pancreatic RNase digestion led to the hypothesis that BDV is a single-stranded RNA virus (DE LA TORRE et al. 1990).

Hybrid arrest experiments suggested that the two classes of cDNA direct the synthesis of two proteins specific for BD, the 40 kDa and the 24 kDa antigens (LIPKIN et al. 1990). Consequently, individual transcripts can be detected in northern hybridization experiments depending on the cDNA probe used: the ORF p40 probe reveals a 1.2 kb signal, while the ORF p24 probe reveals a 0.8 kb signal (Fig. 1). The largest RNA, of approximately 8.5 kb, and the 2.1 kb RNA cross-hybridize with the both probes, whereas the 3.5 kb RNA hybridizes only with the ORF p24 probe. The use of strand-specific RNA and oligonucleotide probes demonstrated a positive polarity for the RNAs of 0.8, 1.2, 2.1 and 3.5 kb (LIPKIN et al. 1990; VANDEWOUDE et al. 1990; Lewis and Lipkin, unpublished). The 0.8, 1.2, 2.1 and 3.5 kb RNAs are polyadenylated (DE LA TORRE et al. 1990; VANDEWOUDE et al. 1990; MCCLURE et al. 1992; THIERER et al. 1992). Whereas an mRNA function for the 0.8 and 1.2 kb RNAs has been established, the role of the 2.1 and 3.5 kb RNAs is not yet clear.

The findings regarding the largest identified RNA species, the putative genomic RNA of BDV, are a subject of controversy. First, VANDEWOUDE et al. (1990)

reported different sizes of 10.5 kb and 8.7 kb for the largest BDV RNA species when derived from BD rat brain or infected MDCK cells, respectively, suggesting that this difference may reflect the presence of a defective genome in the persistently infected cell cultures. DE LA TORRE et al. (1990) on the other hand, found no obvious difference in size for the largest BDV RNA species when derived from BD rat brain or the MBV cell line. Although we initially estimated a size of about 8.5 kb for the largest BDV RNA (LIPKIN et al. 1990; DE LA TORRE et al. 1990; BRIESE et al. 1992), recent results indicate that the size is closer to 10 kb

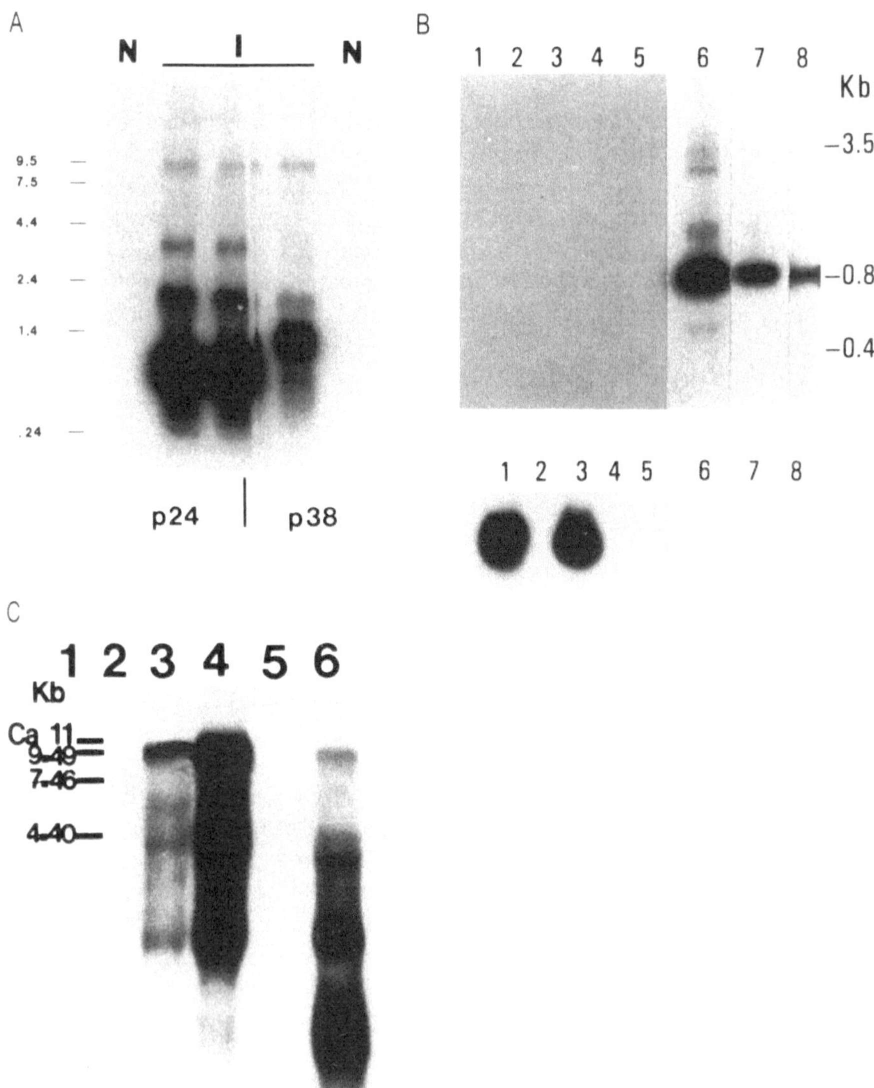

(Fig. 1) (de la Torre, unpublished). Second, VANDEWOUDE et al. (1990) reported the largest RNA as poly A^+ and as being present in equal quantities of positive and negative polarity in BD rat brain extract. In contrast, others have found the largest RNA species to be poly A^- and present predominantly in negative polarity in RNA obtained from BD rat brain or MBV cells (LIPKIN et al. 1990; DE LA TORRE et al. 1990). These discrepancies led to two different models for BDV, with some authors favoring a coronavirus model (VANDEWOUDE et al. 1990; RICHT et al. 1991) and others suggesting that BDV is a negative, single-stranded RNA virus (LIPKIN et al. 1990; DE LA TORRE et al. 1990). The isolation and characterization of RNA contained in the virion was required in order to determine the genome polarity of BDV.

2.2 BDV Contains Genomic RNA of Negative Polarity

Based on earlier observations, that infectious material can be released from BDV-infected tissue culture cells under hypertonic conditions, a method for the isolation of viral particles has been established (PAULI and LUDWIG 1985; BRIESE et al. 1992). Infectious BDV particles were released from tissue culture cells by a buffered 250 m*M* $MgCl_2$ solution and subjected to mild detergent treatment followed by extensive DNase/RNase digestion. Nucleic acid obtained from such released virus preparations revealed only a single RNA of approximately 8.5 kb when probed with the available BDV cDNAs in northern hybridization experiments (Fig. 2). Strand-specific hybridizations using RNA and oligonucleotide probes representing ORF p24 or ORF p40 detected only negative polarity RNA in released virus preparations (Fig. 2) (BRIESE et al. 1992). These results characterize BDV as a negative strand RNA virus; however, additional genomic RNAs not detected by the presently available cDNAs cannot be excluded. Thus, whether BDV has a segmented and/or ambisense genome remains to be determined.

Fig. 1A–C. Characterization of Borna disease virus (BDV) as a RNA virus.
A BDV-specific RNAs. Northern hybridization of total RNA obtained from infected (*I*) and noninfected (*N*) tissue culture cells using ^{32}P-labeled, hexamer-primed DNA probes representing either ORF p40 (p38) or ORF p24 (p24). Positions of RNA markers are indicated by their size in kb. **B** BDV sequences are not present in cellular or episomal DNA (DE LA TORRE et al. 1990). Southern hybridization of *EcoRV*-digested genomic and episomal DNA obtained from MBV and uninfected MDCK cells using a ^{32}P-labeled, hexamer-primed, internal *EcoRV* fragment from the ORF p40 representing cDNA (*top*). As control, the membrane was stripped from the BDV probe and rehybridized to a somatostatin probe (*bottom*). *Lanes 1, 3* genomic DNA (15 μg) from MDCK or MBV cells, respectively; *lanes 2, 4*, episomal DNA (7 μg) from MDCK or MBV cells, respectively; *lane 5* episomal DNA (20 μg) from MBV cells; *lanes 6,7 and 8*, *EcoRV*-digested plasmid DNA containing ORF p40 cDNA, 100, 10 and 1 pg, respectively. Positions of DNA markers are indicated by their size in kb. **C** The largest BDV RNA found in BD rat brain has the same size as the largest one found in BDV-infected tissue culture cells. Northern hybridization using a ^{32}P-labeled, hexamer-primed DNA probe representing ORF p24 and the following RNA samples: *lane 1*, RNA marker ladder (BRL #56205A) containing vesicular stomatitis virus genomic RNA (~11 kb); *lane 2*, total RNA from C6 cells; *lane 3*, nuclear, $polyA^-$ RNA from MBV cells; *lane 4*, nuclear, $polyA^-$ RNA from BDV infected C6 cells; *lanes 5 and 6*, total RNA from uninfected and BD rat brain, respectively. Positions of RNA markers were determined by methylene blue staining of the membrane and are indicated in *lane 1* by their size in kb

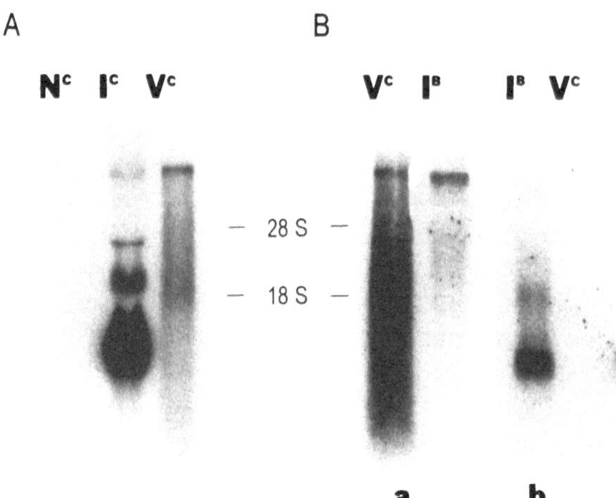

Fig. 2A, B. The largest Borna disease virus (BDV) RNA is genomic and of negative polarity. **A** Released virus preparations contain the 8.5 kb BDV RNA (BRIESE et al. 1992). Northern hybridization of total RNA obtained from uninfected (N^c) and infected (I^c) tissue culture cells or released virus preparation (V^c) using a ^{32}P-labeled, hexamer-primed DNA probe representing ORF p24. **B** The 8.5 kb genomic RNA is of negative polarity (BRIESE et al. 1992). Northern hybridization of total RNA obtained from infected rat brain (I^B) or released virus preparation (V^c) using ^{32}P-labeled, single-stranded RNA probes representing either the antisense strand (*b*) or the sense strand (mRNA orientation) (*a*) of ORF p24. Positions of 18S and 28S rRNAs are indicated

The BDV-specific cDNA clones isolated so far provide a total of about 2 kb of sequence information for BDV (THIERER et al. 1992; MCCLURE et al. 1992). Reverse transcriptase PCR experiments showed that the two ORFs, ORF p40 and ORF p24, are located directly adjacent to each other on the genomic RNA. Amplification of an intervening sequence in these experiments, using a sense ORF p40 primer for reverse transcriptase in conjunction with an antisense ORF p24 primer in PCR, indicated that ORF p24 maps 105 nucleotides 5' with respect to ORF p40 on the negative strand BDV genome (Lipkin, unpublished).

3 Transcription

3.1 Transcription of BDV Occurs in the Nucleus of the Infected Cell

Two observations suggested the possibility that BDV might have a nuclear phase during transcription and/or replication. First, though the 38/40 kDa and the 24 kDa antigens are found in the cytoplasm, they are also present at high levels in the nucleus of infected cells (LUDWIG et al. 1988). Second, the production of these proteins can be prevented by actinomycin D (DUCHALA et al. 1989). More recently,

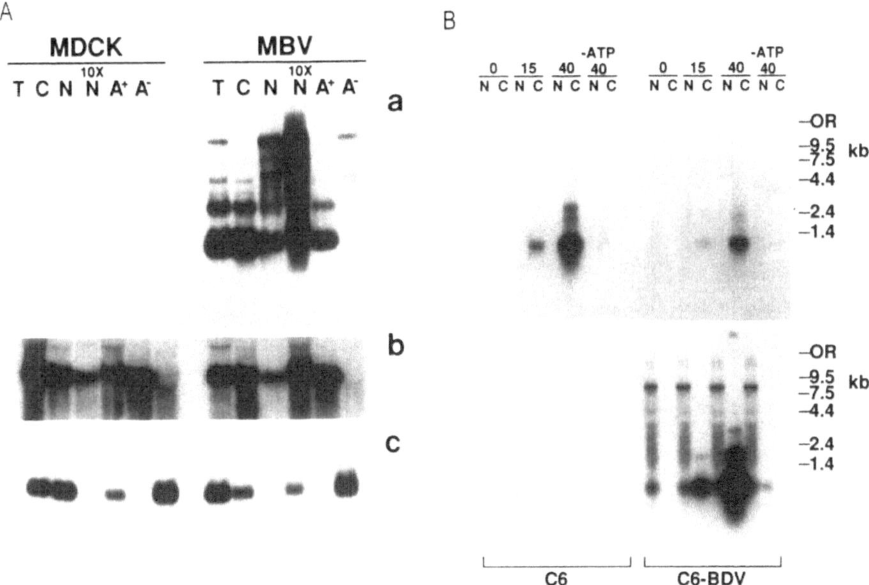

Fig. 3A, B. Poly A⁺ Borna disease virus (BDV) RNAs are transcribed in the nucleus of the infected cell. **A** Cellular distribution and polyadenylation status of BDV RNAs. Northern hybridization of RNA obtained from MDCK (*MDCK*) or MBV (*MBV*) cells after cell fractionation or polyA⁺ selection using ^{32}P-labeled, hexamer-primed DNA probes representing ORF p24 (*a*); vimentin, a polyA⁺ mRNA (*b*) or histone 2b, a non polyadenylated mRNA (*c*). T, RNA obtained from total cells; *C*, RNA obtained from cytoplasmic fraction; *N*, RNA obtained from nuclei; *A⁺*, polyA⁺ RNA; *A⁻*, polyA⁻ RNA. **B** Transcription of BDV takes place in the nucleus of the infected cell (BRIESE et al. 1992). Nuclei from C6 and BDV-infected C6 cells were isolated and RNA nucleocytoplasmic transport assays performed. At indicated time points (0, 15, 40 min) RNA was extracted from nuclear (*N*) and postnuclear (*C*) fractions and analyzed by northern hybridization using ^{32}P-labeled, hexamer-primed DNA probes representing ORF p24 (*bottom*) or the housekeeping gene cyclophilin (*top*). As a control, RNA nucleocytoplasmic transport assays were performed in the absence of ATP (–ATP, 40 min). OR, lane origin

studies of the cellular distribution of BDV RNAs after cell fractionation demonstrated that the largest, polyA⁻ BDV RNA is restricted to the nucleus, whereas the smaller, polyA⁺ RNAs are found in both the nuclear and cytoplasmic fractions (Fig. 3) (DE LA TORRE et al. 1990). The results obtained by cell fractionation were further confirmed by in situ hybridization experiments showing BDV RNAs of positive polarity in a diffuse distributed manner over the entire infected cell, whereas the genomic RNA of negative polarity was confined to the nucleus (CARBONE et al. 1991). To investigate whether transcription of BDV occurs in the nucleus of the infected cell, RNA nucleocytoplasmic transport assays were performed (BRIESE et al. 1992). Nuclei from infected tissue culture cells were isolated, resuspended in nuclear transport buffer and RNA was extracted at various time points from the nuclear and postnuclear fractions and analyzed by northern hybridization using a ORF p24 probe. As shown in Fig. 3, the 0.8, 2.1 and 3.5 kb polyA⁺ RNAs accumulated in the postnuclear fraction in a time-dependent manner. Nucleocytoplasmic transport of these RNAs was prevented in the absence of ATP, indicating a specific, energy-dependent transport of the transcripts. These results indicate that transcription of BDV, a negative, single-stranded RNA virus, takes place in the nucleus of the infected cell (BRIESE et al. 1992).

4 Proteins

4.1 Borna Disease-Specific Proteins

Borna disease is associated with the presence of a soluble antigen (s-antigen) (VON SPROCKHOFF and NITZSCHKE 1955). The s-antigen is found in high quantities in the noninfectious supernatant obtained after high speed centrifugation of sonicated, infected brain or tissue culture cells. Diseased animals have high titers of polyclonal serum antibodies and specific oligoclonal CSF antibodies that are directed against this antigen (LUDWIG et al. 1977; LUDWIG and THEIN 1977; DANNER et al. 1978). The two major components of the s-antigen are the 38/40 kDa and the 24 kDa proteins (LUDWIG et al. 1977, 1988; LUDWIG and BECHT 1977; BAUSE-NIEDRIG et al. 1992). These proteins are not only diffusely distributed in the cytoplasm, but are also present in the nucleus of infected cells, where they appear as circumscribed aggregates (WAGNER et al. 1968; DANNER 1977; LUDWIG and BECHT 1977; BAUSE-NIEDRIG et al. 1991). Colocalization studies, together with the finding that both proteins copurify, indicate that these proteins are likely to form a high molecular weight complex in the cell (BAUSE-NIEDRIG et al. 1991, 1992).

Based on identical immunologic properties, it has been suggested that the 38 kDa and 40 kDa polypeptides are closely related and presumably represent modified forms of the same protein (HAAS et al. 1986; BAUSE-NIEDRIG et al. 1992). The 24 kDa protein, in contrast, appears to be a different, unrelated protein (see also Sect, 4.2). Though monoclonal antibodies reactive with both proteins have been isolated (THIEDEMANN et al. 1992) indicating the presence of one or more common epitopes, polyclonal sera raised against each of the polypeptides reacted only with the 24 kDa or the 38/40 kDa protein but not with both (BAUSE-NIEDRIG et al. 1992). The nature of the modification which accounts for the different M_r of the 38 and 40 kDa polypeptides is still unclear; neither phosphorylation nor glycosylation has been demonstrated for either of the 38 or 40 kDa polypeptides. The 24 kDa protein is phosphorylated (THIEDEMANN et al. 1992).

Immunoblot analysis of s-antigen preparations indicated, in addition to the 38/40 kDa and 24 kDa proteins, the presence of a 60 kDa polypeptide, predominantly found in preparations from BD rabbit brain (LUDWIG et al. 1988). While it is possible that the 60 kDa signal represents an additional, independent protein, recent data indicate that it is likely to represent a multimeric form of p24 that is rather stable towards treatment with reducing agents. Evidence for such a multimeric form of p24 comes from three observations. First, incubation of antigen preparations with

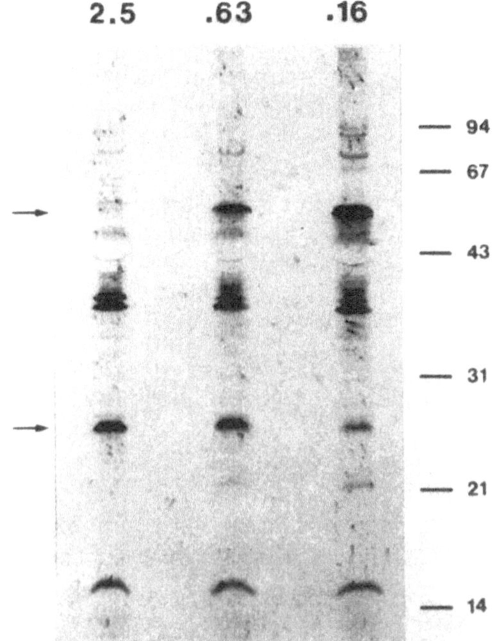

Fig. 4. Borna disease virus (BDV)-specific proteins. The 60 kDa/24 kDa signal is dependent on dithiothreitol (DTT) concentration. Immunoblot of crude protein from $MgCl_2$-released virus using anti-BDV serum and alkaline phosphatase conjugate as second antibody in conjunction with a fast red/naphthol phosphate stain. Concentrations of DTT are indicated on *top* in m*M*. *Arrows* indicate position of 60 kDa and 24 kDa signal. The position of molecular weight markers are indicated by their M_r in kDa

increasing dithiothreitol concentrations resulted in a loss of 60 kDa signal with a concomitant increase in 24 kDa signal (Fig. 4) (Briese and Ludwig, unpublished). Second, purified 24 kDa protein ran either as a 24 kDa or a 60 kDa band depending on the use of reducing or nonreducing conditions in SDS-PAGE, respectively (Briese and Ludwig, unpublished). Third, recombinant 24 kDa protein was also found in a 60 kDa position when nonreducing SDS-PAGE conditions were used (Thibault, Kliche, Briese, and Lipkin, unpublished).

Finally, a proteinase K resistant protein with a M_r of 14.5 kDa has been described (Schädler et al. 1985). The 14.5 kDa protein was isolated from brains of neonatally infected rats using a detergent/salt extraction protocol. Like the 38/40, 24 and 60 kDa antigens, it is exclusively found in infected material, but usually not in s-antigen preparations. The localization of the 14.5 kDa protein in the infected cell is not known.

4.2 Analysis of Cloned Viral Proteins

The two cDNA species cloned from BDV contain potential ORFs for a 39.5 kDa or a 22.5 kDa protein, respectively. Their deduced amino acid sequences are in agreement with partial amino acid sequence obtained through direct sequencing of either purified 40 kDa or purified 24 kDa antigen from BDV-infected material (Thierer et al. 1992; McClure et al. 1992). These data provide conclusive evidence that the cloned viral mRNAs encode two proteins, p40 and p24, which have immunologically been characterized as the 40 kDa and the 24 kDa components of the BD-specific s-antigen. Amino acid sequence comparison reveals no significant homology between p24 and p40. In hybrid arrest experiments translation of both the 38 kDa and 40 kDa antigen is blocked by cDNA representing ORF p40 (LIPKIN et al. 1990). In bacterial expression systems ORF p40 directs translation of a product that runs as a double band with a M_r of 38/40 kDa in SDS-PAGE (Lipkin, unpublished). These results, in conjunction with the immunologic data, demonstrate a close relationship for these two polypeptides and suggest that the 38 kDa antigen represents a modified version of p40. Presently, we have no mechanism to explain how the ORF p40 directs translation of both the 38 kDa and 40 kDa polypeptides.

Database searches for similarities with other reported sequences have indicated a distant similarity of p40 to a duplicated domain of L-polymerases from paramyxo- and rhabdoviruses (McCLURE et al. 1992). Paramyxo- and rhabdoviruses are members of the order *Mononegavirales*, enveloped RNA viruses with nonsegmented genomes of negative polarity, to which BDV may be related. However, this similarity concerns one domain of a single protein. L-polymerases are multifunctional enzymes more that 200 kDa in size and are synthesized in low abundance. In contrast, p40 is much smaller and is present in much higher amounts than would be anticipated for a functional polymerase enzyme. Nonetheless, p40 is present in the nucleus of the infected cell and might act in concert with p24, which shares the nuclear location. Colocalization studies using monoclonal antibodies against the two proteins have indicated that both proteins are present in a high molecular weight complex found in the nucleus of infected cells (BAUSE-NIEDRIG et al. 1991). In accordance with their nuclear localization, putative nuclear localization signals are found in both proteins. For p40 the positively charged sequence $K^{149}KRFK^{153}$ has been proposed as a nuclear targeting signal, whereas for p24 the basic amino acid cluster $R^{22}RKRSGSPRPRK^{33}$ has been proposed to serve this function because of its similarity to a sequence motif shown to determine the nuclear localization of hepatitis B virus core antigen (McCLURE et al. 1992; THIERER et al. 1992).

The accumulation of high amounts of p40 and p24 in infected cells could also be compatible with a function of the proteins as structural components of BDV. Even their nuclear localization would be consistent with a structural role analogous to the nucleoprotein in the influenza virus system. If present in the virion, then p40 and p24 are unlikely to be exposed at the surface, because no labeling of BDV particles could be achieved with antibodies against p40 or p24 (Lipkin and Ribak, unpublished). Further biochemical studies will be needed to assign any function to these proteins.

5 Concluding Remarks

Though the body of data accumulated over the past 3 years has led to consensus on the characterization of BDV as a RNA virus, some aspects of BDV's biology are still controversial.

VANDEWOUDE, RICHT and coworkers had suggested that the BD-specific 14.5 kDa protein is a virus-encoded protein expressed from a smaller ORF present within the cDNA clone containing ORF p24 that overlaps with ORF p24 (VANDE WOUDE et al. 1990; RICHT et al. 1991). However, partial amino acid sequence from BD-specific 14.5 kDa protein purified following the method of SCHÄDLER et al. (1985) (NSKHSYV; KLICHE et al. 1994) does not match the amino acid sequence deduced from the ORF proposed by RICHT, VANDEWOUDE and coworkers (RICHT et al. 1991). Instead, the amino acid data identified a potential ORF p16, residing adjacent to the p24 gene in nucleic acid sequence obtained from genomic BDV cDNA clones. This suggests that the 14.5 kDa protein is a viral protein encoded by a gene separate from ORF p24 (KLICHE et al. 1994).

Table 1. Characterization of Borna disease virus

Nucleic acids
Negative, single-stranded genomic RNA
8.5–10.5 kb
poly (A)$^+$ transcripts transcribed in the nucleus
0.8 kb p24 mRNA
1.2 kb p40/38 mRNA
2.1 kb (precursor, 1.2 kb + 0.8 kb)?
3.5 kb (precursor, 0.8 kb + ?)?
Proteins
38/40 kDa protein
Virus-encoded
Component of s-antigen (complex)
Cytoplasmic and nuclear localization
Putative nuclear localization signal
Distant similarity to polymerases of paramyxo- and rhabdoviruses
Function unknown
24 kDa protein
Virus-encoded
Component of s-antigen (complex)
Cytoplasmic and nuclear localization
Putative nuclear localization signal
Ability to form 60 kDa multimer (dimer?)
Function unknown
14 kDa protein
Virus-encoded
Proteinase K resistant
Function unknown
60 kDa antigen
Component of s-antigen (complex)
(Multimeric form of p24?)
Function unknown
Morphology
(Enveloped, spherical 90 nm particle)

In the same publications, RICHT, VANDEWOUDE and coworkers have discussed the occurrence of a nested set of overlapping subgenomic, positive- and negative-strand RNAs in BDV, proposing a similarity of BDV to coronaviruses, which synthesize nested sets of overlapping subgenomic mRNAs that are 3'-coterminal with the genome (VANDEWOUDE et al. 1990; RICHT et al. 1991). However, Fig. 1 shows that BDV synthesizes several transcripts that are not coterminal, which argues against a coronavirus model.

Table 1 summarizes the present knowledge of the molecular biology of BDV. BDV is a single-stranded RNA virus containing a genomic RNA of 8.5–10.5 kb. This genomic RNA is of negative polarity for sequences complementary to the two mRNAs presently cloned from BDV; no data are available to decide whether the BDV genome is segmented and/or ambisense. Four polyA$^+$ transcripts of 3.5, 2.1,1.2 and 0.8 kb are identified by the available BDV cDNAs. Transcription of the RNAs takes place in the nucleus as determined by RNA nucleocytoplasmic transport assays. The role of the 2.1 and 3.5 kb BDV transcripts is unclear. Based on its size, the 2.1 kb RNA could represent a dicistronic RNA of ORF p40 plus ORF p24; the 3.5 kb RNA might contain other, additional gene(s). These RNAs may represent read-through artifacts or function as precursor molecules which are processed into individual mRNAs. BDV mRNAs of 1.2 kb and 0.8 kb direct the translation of p40 and p24, respectively, two viral proteins that have previously been characterized as the major components of the BD-specific s-antigen. Both proteins are found in the cytosol and in the nucleus of infected cells and have motifs consistent with nuclear localization signals. Comparative sequence analysis has revealed a distant similarity of p40 to L-polymerases of paramyxo- and rhabdoviruses. In addition to p40 and p24, BDV also encodes p16, a protein described as the BD-specific 14.5 kDa protein.

BDV still remains unclassified. Among RNA viruses only orthomyxoviruses, which have a negative-stranded, segmented genome, are known to transcribe in the nucleus of the infected cell (HERZ et al. 1981; JACKSON et al. 1982). Thus, if a nonsegmented genome is confirmed, BDV might be considered to represent a new class of virus, and it would be conceivable that other BDV related viruses will be found.

Identification of a new BDV-related class of viruses might have potential implications for human disease: immunologic data indicate that BDV/BDV-related viruses may infect humans (see Bode, this volume), and some manifestations of BD resemble human neuropsychiatric disorders of unknown cause (see Solbrig, Fallon and Lipkin, this volume). Studies into the molecular genetics of BDV now provide sensitive probes to evaluate the possible involvement of BDV/BDV-related viruses in human CNS disorders. Evidence that viruses can induce progressive neurological disorders has encouraged studies to understand the mechanisms and biologic consequences of persistent virus infection of the CNS (GILDEN and LIPTON 1989; TER MEULEN 1991). In this regard, BDV has become an attractive model for the investigation, not only of immune-mediated pathological events in virally induced neurological disease (Stitz, Dietzschold and Carbone, this volume), but also of the mechanisms by which virus infection of the CNS can lead to subtle disturbances in behavior (SPRANKEL et al. 1978; DITTRICH et al. 1989).

Further, understanding of the molecular genetics and biology of BDV offers the opportunity to characterize the basis for its unique tropism. The unique aspects of BDV tropism anticipate its importance for neurobiology and clinical medicine. Studies are already underway to use BDV as a tool for mapping connectivity within the nervous system. As the molecular basis for pathogenicity becomes clear, it may be feasible to establish BDV-based vectors for targeted drug delivery within the brain.

Acknowledgements: Support for this work was provided by NIH grants NS-29425 (T.B., W.I.L.); NS-12428 (J.C.T.); UC Taskforce on AIDS-R911047 (T.B.,W.I.L.). T.B. is a recipient of a Young Investigator Award from NARSAD. W.I.L. is recipient of a Young Investigator Award from NARSAD and a Pew Scholars Award from the Pew Charitable Trusts.

References

Amsterdam JD, Winokur A, Dyson W, Herzog S, Gonzalez F, Rott R, Koprowski H (1985) Borna disease virus. A possible etiologic factor in human affective disorders? Arch Gen Psychiat 42: 1093–1096
Anzil AP, Blinzinger K, Mayr A (1973) Persistent Borna virus infection in adult hamsters. Arch Gesamte Virusforsch 40: 52–57
Bause-Niedrig I, Pauli G, Ludwig H (1991) Borna disease virus-specific antigens: two different proteins identified by monoclonal antibodies. Vet Immunol Immunopathol 27: 293–301
Bause-Niedrig I, Jackson M, Schein E, Ludwig H, Pauli G (1992) Borna disease virus-specific antigens. II. The solube antigen is a protein complex. Vet Immunol Immunopathol 31: 361–369
Bode L, Riegel S, Ludwig H, Amsterdam JD, Lange W, Koprowski H (1988) Borna disease virus-specific antibodies in patients with HIV infection and with mental disorders. Lancet ii: 689
Bode L, Riegel S, Lange W, Ludwig H (1992) Human infections with Borna disease virus: seroprevalence in patients with chronic diseases and healthy individuals. J Med Virol 36: 309–315
Briese T, de la Torre JC, Lewis A, Ludwig H, Lipkin WI (1992) Borna disease virus, a negative-strand RNA virus, transcribes in the nucleus of infected cells. Proc. Natl. Acad. Sci. USA 89: 11486–11489
Briese T, Schneemann A, Lewis AJ, Park YS, Kim S, Ludwig H, Lipkin WI (1994) Genomic organization of Borna disease virus. Proc Natl Acad Sci USA 91: 4362–4366
Carbone KM, Duchala CS, Griffin JW, Kincaid AL, Narayan O (1987) Pathogenesis of Borna disease in rats: evidence that intra-axonal spread is the major route for virus dissemination and the determinant for disease incubation. J. Virol. 61: 3431–3440
Carbone KM, Moench TR, Lipkin WI (1991) Borna disease virus replicates in astrocytes, schwann cells and ependymal cells in persistently infected rats: location of viral genomic and messenger RNAs by in situ hybridization. J. Neuropathol. Exp. Neurol. 50: 205–214
Cubitt B, Oldstone C, de la Torre JC (1994) Sequence and genome organization of Borna disease virus. J Virol 68: 1382–1396
Cubitt B, Oldstone C, Valcarel V, de la Torre JC (1994) RNA splicing contributes to the generation of mature mRNAs of Borna disease virus, a non-segmented negative strand RNA virus. Virus Res 34: 69–79
Cubitt B, de la Torre JC (1994) Borna disease virus (BDV), a nonsegmented RNA virus, replicates in nuclei of infected cells where infectious BDV ribonucleoproteins are present. J Virol 68: 1371–1381
Compans RW, Melsen LR, de la Torre JC (1994) Virus-like particles in MDCK cells persistently infected with Borna disease virus. Virus Res 33: 261–268
Danner K (1977) Borna disease: patterns of infection. In: ter Meulen V, Katz M (eds) Slow virus infections of the central nervous system. Springer, Berlin Heidelberg New York, pp 84–88
Danner K, Mayr A (1979) *In vitro* studies on borna virus. II. Properties of the virus. Arch. Virol. 61: 261–271
Danner K, Lüthgen K, Herlyn M, Mayr A (1978) Vergleichende Untersuchungen über Nachweis und Bildung von Serumantikörpern gegen das Borna-Virus (in German). Zentralbl Vet Med [B] 25: 345–355
de la Torre JC, Carbone KM, Lipkin WI (1990) Molecular characterization of the borna disease agent. Virology 179: 853–856
Dittrich W, Bode L, Ludwig H, Kao M, Schneider K (1989) Learning deficiencies in Borna disease virus-infected but clinically healthy rats. Biol Psychiatry 26: 818–828

Duchala CS, Carbone KM, Narayan O (1989) Preliminary studies on the biology of Borna disease virus. J Gen Virol 70: 3507–3511

Elford WJ, Galloway IA (1933) Filtration of the virus of Borna disease through graded collodion membranes. Br J Exp Pathol 14: 196–205

Gilden DH, Lipton HL (eds) (1989). Clinical and molecular aspects of neurotropic virus infection. Kluwer Academic, Boston

Gosztonyi G, Ludwig H (1984) Borna disease of horses. An immunohistological and virological study of naturally infected animals. Acta Neuropathol (Berl) 64: 213–221

Haas B, Becht H, Rott R (1986) Purification and properties of an intranuclear virus-specific antigen from tissue infected with borna disease virus. J Gen Virol 67: 235–241

Heinig A (1969) Die Bornasche Krankheit der Pferde und Schafe (in German). In: Röhrer H (ed) Handbuch der Virusinfektionen bei Tieren, Vol 4. Fischer, Jena, pp 83–148.

Herz C,Stavnezer E, Krug R, Gurney T (1981) Influenza virus, an RNA virus, synthesizes its messenger RNA in the nucleus of infected cells. Cell 26: 391–400

Hirano N, Kao M, Ludwig H (1983) Persistent, tolerant or subacute infection in Borna disease virus-infected rats. J Gen Virol 64: 1521–1530

Jackson DA, Caton AJ, McCready SJ, Cook PR (1982) Influenza virus RNA is synthesized at fixed sites in the nucleus. Nature 296: 366–368

Joest E, Degen K (1911) Untersuchungen über die pathologische Histologie, Pathogenese und postmortale Diagnose der seuchenhaften Gehirn-Rückenmarksentzündung (Bornasche Krankheit) des Pferdes (in German). Z Inf Krkh Haustiere 9: 1–98

Kao M, Ludwig H, Gosztonyi G (1984) Adapation of borna disease virus to the mouse. J Gen Virol 65: 1845–1849

Kliche S, Briese T, Henschen AH, Stitz L, Lipkin WI (1994) Characterization of a Borna disease virus glycoprotein, gp18. J Virol 68: 6918–6923

Krey H, Ludwig H, Rott R (1979) Spread of infectious virus along the optic nerve into the retina in Borna disease virus-infected rabbits. Arch Virol 61: 283–288

Lipkin WI, Travis GH, Carbone KM, Wilson MC (1990) Isolation and characterization of Borna disease agent cDNA clones. Proc Natl Acad Sci USA 87: 4184–4188

Ludwig H, Becht H (1977) Borna, the disease—a summary of our present knowledge. In: ter Meulen V, Katz M (eds) Slow virus infections of the central nervous system. Springer, Berlin Heidelberg New York, pp 75–83

Ludwig H, Thein P (1977) Demonstration of specific antibodies in the central nervous system of horses naturally infected with Borna disease virus. Med Microbiol Immunol 163: 215–226

Ludwig H, Becht H, Groh L (1973) Borna disease (BD), a slow virus infection. Biological properties of the virus. Med Microbiol Immunol 158: 275–289

Ludwig H, Koester V, Pauli G, Rott R (1977) The cerebrospinal fluid of rabbits infected with borna disease virus. Arch Virol 55: 209–223

Ludwig H, Kraft W, Kao M, Gosztonyi G, Dahme E, Krey H (1985) Borna-Virus-Infektion (Borna-Krankheit) bei natürlich und experimentell infizierten Tieren: ihre Bedeutung für Forschung und Praxis (in German). Tierarztl Prax 13: 421–453

Ludwig H, Bode L, Gosztonyi G (1988) Borna disease: a persistent virus infection of the central nervous system. Prog Med Virol 35: 107–151

Matthias D (1954) Der Nachweis von latent infizierten Pferden, Schafen und Rindern und deren Bedeutung als Virusreservoir bei der Bornaschen Krankheit (in German). Arch Exp Vet Med 8: 506–511

McClure MA, Thibault KJ, Hatalski CG, Lipkin WI (1992) Sequence similarity between Borna disease virus p40 and a duplicated domain within the paramyxovirus and rhabdovirus polymerase proteins. J Virol 66: 6572–6577

Metzler A, Frei U, Danner K (1976) Virologisch gesicherter Ausbruch der Bornashen Krankheit in einer Schafherde der Schweiz (in German). Schweiz Arch Tierheilkd 118: 483–492

Morales JA, Herzog S, Kompter C, Frese K, Rott R (1988) Axonal transport of Borna disease virus along olfactory pathways in spontaneously and experimentally infected rats. Med Microbiol Immunol 177: 51–68

Narayan O, Herzog S, Frese K, Scheefers H, Rott R (1983) Pathogenesis of Borna disease in rats: immune-mediated viral ophthalmoencephalopathy causing blindness and behavioral abnormalities. J Infect Dis 148: 305–315

Nicolau S, Galloway IA (1928) Borna disease and enzootic encephalo-myelitis of sheep and cattle. Stationary Office, London (Privy Council, Medical Research Council, special report series no 121)

Nitzschke E (1963) Untersuchungen über die experimentelle Bornavirus-Infektion bei der Ratte (in German). Zentralbl Vet Med [B] 10: 470–527

Pauli G, Ludwig H (1985) Increase of virus yields and releases of Borna disease virus from persistently infected cells. Virus Res. 2: 29–33.

Pauli G, Grunmach J, Ludwig H (1984) Focus-immunoassay for borna disease virus-specific antigens. Zentralbl Vet Med [B] 31: 552–557

Richt JA, VandeWoude S, Zink MC, Narayan O, Clements JE (1991) Analysis of Borna disease virus-specific RNAs in infected cells and tissues. J Gen Virol 72: 2251–2255

Richt JA, VandeWoude S, Zink MC, Clements JE, Herzog S, Stitz L, Rott R, Narayan O (1992) Infection with Borna disease virus: molecular and immunobiological characterization of the agent. Clin. Infect Dis 14: 1240–1250

Rott R, Herzog S, Fleischer B, Winokur A, Amsterdam J, Dyson W, Koprowski H (1985) Detection of serum antibodies to Borna disease virus in patients with psychiatric disorders. Science 228: 755–756

Rott R, Herzog S, Bechter K, Frese K (1991) Borna disease, a possible hazard for man? Arch Virol 118: 143–149

Schädler R, Diringer H, Ludwig H (1985) Isolation and characterization of a 14500 molecular weight protein from brains and tissue cultures persistently infected with Borna disease virus. J Gen Virol 66: 2479–2484

Schneeman A, Schneider PA, Kim S, Lipkin WI (1994) Identification od signal sequences that cintrol transcription of Borna disease virus, a nonsegmented, negative-strand RNA virus. J Virol 68: 6514–6522

Schneider S, Breise T, Henschen AH, Stitz L, Lipkin WI (1994) Characterization of a Borna disease virus glycoprotein, gp18. J Virol 68: 6918–6923

Seifried O, Spatz H (1930) Die Ausbreitung der encephalitischen Reaktion bei der Bornaschen Krankheit der Pferde und deren Beziehungen zu der Encephalitis epidemica, der Heine-Medinschen Krankheit und der Lyssa des Menschen. Eine vergleichend-pathologische Studie (in German). Z Gesamte Neurol Psychiatry 124: 317–382

Sprankel H, Richarz K, Ludwig H, Rott R (1978) Behavior alterations in tree shrews (Tupaia glis, Diard 1820) induced by Borna diseasae virus. Med Microbiol Immunol 165: 1–18

Stitz L, Krey H, Ludwig H (1980) Borna disease in Rhesus monkeys as a model for uveo-cerebral symptoms. J Med Virol 6: 333–340

ter Meulen V (ed) (1991) Virus-cell interactions in the nervous system. Seminars in neuroscience, Vol 3. London, Saunders, pp 81–123

Thiedemann N, Presek P, Rott R, Stitz L (1992) Antigenic relationship and further characterization of two major Borna disease virus-specific proteins. J Gen Virol 73: 1057–1064

Thierer J, Riehle H, Grebenstein O, Binz T, Herzog S, Thiedemann N, Stitz L, Rott R, Lottspeich F, Niemann H (1992) The 24K protein of Borna disease virus. J Gen Virol 73: 413–416.

VandeWoude S, Richt JA, Zink MC, Rott R, Narayan O, Clements JE (1990) A Borna virus cDNA encoding a protein recognized by antibodies in humans with behavioral diseases. Science 250: 1278–1281

von Sprockhoff H (1956) Zur biologischen Charakterisierung des Borna-s-Antigens (in German). Z. Immun Forsch (Jena) 113: 379–385

von Sprockhoff H, Nitzschke E (1955) Untersuchungen über das komplementbindende Antigen in Gehirnen bornavirus-infizierter Kaninchen. 1. Mitteilung: Nachweis eines löslichen Antigens (in German). Zentralbl Vet Med 2. 185–192

Waelchli RO, Ehrensperger F, Metzler A, Winder C (1985) Borna disease in a sheep. Vet Rec 117: 499–500

Wagner K, Ludwig H, Paulsen J (1968) Fluoreszenzserologischer Nachweis von Borna-Virus Antigen (in German). Berl Munch Tierarztl Wochenschr 81: 395–396

Zimmermann W, Breter H, Rudolph M, Ludwig H (1994) Borna disease virus: immunoelectron microscopic characterization of cell-free virus and further information about the genome. J Virol 68: 6755–6758

Zwick W (1939) Bornasche Krankheit und Enzephalomyelitis der Tiere (in German). In: Gildemeister E, Haagen E, Waldmann O (eds) Handbuch der Viruskrankheiten, vol 2. Fischer, Jena, pp 254–354

Zwick W, Seifried O, Witte J (1926) Experimentelle Untersuchungen über die seuchenhafte Gehirn- und Rückenmarksentzündung der Pferde (Bornasche Krankheit) (in German). Z Inf Krankh Haustiere 30: 42–136

Note Added in Proof

Since this chapter was written, the BDV genome has been cloned and sequenced, subgenomic RNAs have been mapped to the viral genome and viral particles have been identified by immunoelectron microscopy. BDV has a nonsegment 8.9 kb genome with complementary 3' and 5' termini that contains antisense information of five ORFs (Briese et al. 1994; Cubitt and de la Torre 1994). Unlike genomes of other nonsegmented, negative strand RNA viruses, the BDV genome does not have distinct gene boundaries. Instead, transcription units and transcription signals frequently overlap (Schneemann et al. 1994). Subgenomic RNAs encoding p40 and p24 are monocistronic. In contrast, RNAs encoding the putative matrix, glycoprotein and polymerase proteins are polycistronic and undergo posttranscriptional modification by RNA splicing (Briese et al. 1994; Cubitt et al. 1994; Schneemann et al. 1994; Schneider et al. 1994). Morphologically the virion appears to be a 90 nm enveloped, spherical particle containing an electron dense internal structure consisting of strand-like material (Zimmermann et al. 1994; Compans et al. 1994). In concert, these features indicate that BDV represents the prototype of a new group of animal RNA viruses within the order *Mononegavirales*.

Natural and Experimental Borna Disease in Animals

R. ROTT and H. BECHT

1 Introduction 17
2 Etiology 18
3 Virus Replication 18
4 Natural Infection 19
4.1 Transmission and Natural Host Spectrum 19
4.2 Pathogenesis 20
4.3 Lesions 20
4.4 Clinical Symptoms 21
4.5 Inapparent Infections 22
5 Experimental Transmission of BDV 23
5.1 Experimental Infection of Rats 24
5.2 Variability of BDV 25
5.3 Infection of Immunodeficient Rats 26
6 Nonsusceptible Animals 26
7 Conclusions 27
References 27

1 Introduction

Borna disease (BD) is a transmissible, progressive polioencephalomyelitis of horses and sheep, which are the main natural hosts. It occurs sporadically in endemic areas of Germany and Switzerland, while its presence in other countries has not been fully substantiated.

The disease owes its name to the area around the town Borna in Saxony, Germany, where a great number of horses died during an epidemic in 1885. Cases of equine encephalitis with symptoms closely resembling the neurological signs of BD had been described under various designations, such as *hitzige Kopfkrankheit*; since the end of the eighteenth century, particularly in southern Germany. Around the turn of this century, studies concerning the pathology and pathophysiology of the disease became of foremost interest (JOEST and DEGEN 1911). Milestones in BD-related research were: (1) the final proof that a

---

Institut für Virologie, Justus-Liebig-Universität Gießen, 35392 Gießen, Germany

neurotropic virus was the etiological agent, by successful transmission of brain homogenates of infected horses to experimental animals (ZWICK and SEIFRIED 1925; ZWICK 1939); (2) the demonstration of virus growth in cell cultures (MAYR and DANNER 1972; LUDWIG et al. 1973; HERZOG and ROTT 1980); and (3) the finding that the pathogenesis of BD is mediated by a T cell-dependent immune mechanism (NARAYAN et al. 1983). Data obtained most recently have shed some light on the nature of the causative virus (see Briese, this volume). Further details on the historical background of BD can be found in a comprehensive review by HIEPE (1958/1959); various aspects of the disease and the causative agent were recently reviewed by RICHT et al. (1992).

2 Etiology

Borna disease virus (BDV) has not been fully characterized; thus, BDV is listed among the unclassified viruses. According to filtration studies, the size of the virus is 85–125 nm (ELFORT and GALLOWAY 1933). Its sensitivity towards lipid solvents like ether, chloroform or acetone indicated that the virus is enveloped. Accordingly, a buoyant density of 1.18–1.22 g/ml was estimated in peak fractions of viral infectivity (LUDWIG et al. 1988). At least three virus-specific proteins with respective molecular masses of 14, 24 and 38/39 kDa were found in infected brain tissue and cell culture (SCHÄDLER et al. 1985; HAAS et al. 1986; STITZ et al. 1991; THIEDEMANN et al. 1992). Most recently, the viral genome has been identified as a single-stranded RNA with negative polarity (see Briese et al., this volume). Viral infectivity is inactivated by UV light at the same rate as in conventional viruses. The virus resists pH values between 5 and 12 and is inactivated by heating at 56°C for 30 min.

3 Virus Replication

Borna disease virus is strictly neurotropic and is disseminated by intra-axonal transport from the site of infection (KREY et al. 1976; MORALES et al. 1988; CARBONE et al. 1987). In intracerebrally (i.c.) or intranasally (i.n.) infected animals, the virus appears after a latent period of 3–5 days in the brain, cerebrospinal fluid (CSF) and adrenal glands. Less commonly, virus can be isolated from the salivary gland, mammary gland, nasal mucous membrane and kidney. In rats, the virus replicates readily in cerebral tissue, reaching plateau titers of about 10^6 $TCID_{50}$/g 10 days after infection (NARAYAN et al. 1983). Intracellular viral antigen can be demonstrated in the cell nucleus by immunohistological methods; staining is less brilliant in the cytoplasm of the infected cells. Viral RNA can be located by in situ hybridization primarily in the nucleus of those cells in which antigens accumulate (CARBONE et al. 1991).

The virus replicates in vitro in embryonic brain cells of various species including neural cells, astrocytes and oligodendrocytes. Cocultivation of such cells with various cell lines, such as MDCK cells or Vero cells, results in a persistent infection. In cell cultures the virus spreads by cell to cell contact. Infectivity remains cell-associated and only minimal infectivity is found in the culture medium. The virus is noncytopathic. Conventional virus particles with a diameter of 60–90 nm have been suspected to represent the causative virus (Richt et al. 1993; Zimmermann et al. 1994). Viral antigen can readily be demonstrated mainly in the cell nucleus (Herzog and Rott 1980). Treatment of infected cells by α/β interferon does not influence the establishment or maintenance of the persistent state (von Rheinbaben et al. 1985).

4 Natural Infection

4.1 Transmission and Natural Host Spectrum

There is general agreement that the virus is transmitted through saliva and nasal secretions. Animals become infected by direct contact with secretions or by exposure to contaminated food or water. It is likely that the nose is the main site of virus entrance into the body. Colostrum and milk have been incriminated to play a role in infection of foals. Contact experiments demonstrated that horses without overt disease (virus carriers) may represent a source of infection (Heinig 1969). This observation is of eminent importance for the introduction of the virus into stables without a previous history of BD. Besides the horse as the prominent natural host, sheep are susceptible to natural infections. According to more recent observations, cattle, rabbits, goats, deer, llamas, alpacas, cats, and ostriches can also have natural infections.

The identity of the etiological agent of BD in *horses* and in *sheep* was demonstrated as early as 1926 by Beck and Froboese and was underlined by the observation that BDV could be transmitted between *horses* and *sheep* in endemic areas. While in horse stables only individual animals usually show clinical symptoms, the disease can affect a large number of animals in a flock of sheep.

Natural infections in *rabbits* were observed more recently. A total of ten cases were diagnosed by clinical and pathological examinations, and the diagnosis was verified by direct demonstration of BDV in brain tissue (Otta and Jentsch 1960; Metzler et al. 1978).

Four cases of clinically manifest cases of BD were registered in *cattle*; three animals originated from Saxony, one case was diagnosed in Switzerland in 1992 (Ernst and Hahn 1927; Caplazi et al. 1994). Diagnosis was established by histopathology, immunohistology and by demonstrating BDV-specific antibodies in blood serum.

Neurological symptoms resembling those of BD in sheep were observed in three *goats*, and typical lesions could be demonstrated in the CNS. One case in *deer* concerned an animal which became conspicuous to a hunter by its unconventional behavior (Ernst and Hahn 1927).

Altmann and coworkers (1976) reported BD in *donkeys, mules*, two *llamas* and two *alpacas*.

More recently evidence was obtained that *cats* and *ostriches* could naturally be infected by BDV (Lundgren et al. 1993; Malkinson et al. 1993).

Finally, it should be emphasized that there is evidence that infections by BDV may represent a health hazard for humans (for a review see Rott et al. 1991; see Bode, this volume).

4.2 Pathogenesis

The virus probably gains access to the CNS by intra-axonal migration through the olfactory nerve. This route of migration could be traced by immunohistological staining of viral antigen in the nucleus of neurons, in the perikaryon, in dendrites and in axons. Other cerebral nerve endings terminating in mucous membranes of the oropharyngeal region may also be involved in the transport of the virus to the brain (Morales et al. 1988). Effective oral infections could mean that nerve endings innervating intestinal regions are also suitable for the uptake of the virus and its transport to the CNS (Heinig 1969).

The virus spreads through the CNS by intra-axonal transport and possibly via the CSF. It localizes preferentially in certain parts of the brain, gray matter, nucleus niger, hippocampus or olfactory bulb. Centrifugal spread can occur from the CNS to peripheral nerves, whereby the virus can reach the ganglia of some organs. This mode of viral spread explains the inflammatory and degenerative alterations in peripheral nerves. Certain focal symptoms are induced by the tropism of the virus for certain regions of the brain; the nucleus niger may explain the appearance of motor disorders. For example, the predilection of BDV for BDV-specific antibodies can be demonstrated in serum and in the CSF of most animals after natural or experimental infections. Antibody titers are mostly relatively low, and the antibodies are neutralizing (Hirano et al. 1983; Ludwig et al. 1988). This report, however, was not supported by others (Narayan et al. 1983), as coexistence of virus-specific antibodies and infectious virus in the CSF contradicts this assumption. Furthermore, the lack of any protective effect for susceptible animals after transfer of virus-specific antibodies argues against the neutralizing capacity of the antibodies. All data available so far indicate that an immune response to viral antigens does not elicit protective immunity but rather an immunopathological reaction in which T cells play the essential role (see Stitz et al. this volume).

4.3 Lesions

Gross lesions are not apparent in the CNS. In some cases slight hyperemia and increased moisture of the brain can be observed; gray-reddish discoloration of the gray matter with scattered minute subependymal petechiae, particularly in the nucleus caudatus, may be discernible in the so-called hemorrhagic form of BD.

The histopathological picture of spontaneous BD in the horse, which was described in detail by JOEST and DEGEN (1911) and SEIFRIED and SPATZ (1930), corresponds to a nonpurulent polioencephalomyelitis. Prominent findings are massive perivascular infiltrations consisting of lymphocytes, plasma cells and monocytes which can penetrate into surrounding cerebral tissue. A reactive astrocytosis occurs regularly. In neurons, and more rarely in glia cells, acidophilic inclusions (JOEST-DEGEN inclusion bodies) can frequently be observed. Demyelinization does not take place (GOSZTONYI and LUDWIG 1984). If the disease runs a prolonged course, a discrete internal hydrocephalus may develop. Inflammatory reactions are prominent in regions of the limbic system and in periventricular regions of the inter-brain and mid-brain, and to a lesser degree in the brainstem; the cerebellum remains unaffected. The spinal cord and roots of spinal nerves and the retina are inconsistently affected in spontaneous cases of BD in the horse. In about 50% of diseased horses a lymphocytopenia can be noted. (Pathology is reviewed in detail by Gosztonyi and Ludwig, this volume.)

4.4 Clinical Symptoms

A minimum incubation period of 4 weeks is estimated for horses and sheep. During the initial stage of clinical manifestations, nonspecific signs can be observed, including hyperthermia up to 38.5°C, anorexia, jaundice, difficulties in swallowing, constipation and colic. Horses may not lay down and may show either excited or depressed behaviors.

The clinical picture in acute disease is dominated by symptoms resulting from meningitis and encephalomyelitis, such as severe depression, permanent standing upright in unphysiological positions, ataxia, breaking down unexpectedly, circular movements and running against obstacles. The animals sometimes "forget" to continue chewing and keep the head stretched forward. Ocular disorders, caused by a lymphocytic retinitis and inflammation of the optic nerve, have been noted. The almost regular onset of paralytic symptoms is accompanied by a general fatigue and frequent bending of joints. During the final stage of disease, some animals assume a decubitus posture and carry out paddling movements with their legs. Respiration during this stage can correspond to the Cheyne-Stokes type. The period of clinical illness is mostly 1–3 weeks, only rarely 1–6 days or more than 3 weeks. Mortality rates of horses range between 80% and 100%; for sheep mortality rates are 50% and higher.

Clinical manifestations of BD may vary in horses and in sheep. In sheep, during the initial stages and sometimes in the later stages of BD, cerebral or bulbar and spinal disorders may dominate the clinical picture. In less severe cases these manifestations can progressively improve; however, permanent defects may remain, which can range from motor difficulties to disturbed behavior. Recurrent episodes are possible, particularly after exposure to stress situations (HEINIG 1969).

BD tends to occur between March and July with a peak incidence in May. The assumption of some authors, that waves of outbreaks follow each other within 2–3 years, has not been substantiated. There is good evidence, however, that the occurrence of BD is significantly more frequent in some years than in others.

Although no definite proof exists, it is plausible that rodents may represent a virus reservoir. In addition, the coincidence of the seasonal appearance of BD with the active period of arthropods could mean that insects play a role as vectors. In Europe, the virus has never been isolated from biting insects. In the Near East, however, ticks were held responsible for the transfer of a form of equine encephalitis which runs a clinical course and presents histopathological lesions indistinguishable from BD (Daubney and Mahlau 1967). Final proof that these outbreaks correspond to BD must await confirmation from a comparative analysis of the causative virus.

4.5 Inapparent Infections

Recent sero-epidemiological studies demonstrated that natural BDV infections occur more frequently in Germany than has previously been assumed. In addition there is evidence that BDV is no longer limited to the known endemic areas. Many seropositive horses turned out to be infected subclinically; the only conspicuous feature being the relatively frequent reports of attacks of colic (Lange et al. 1987).

Seropositive horses without clinical manifestations were not only detected in stables with previous cases of BD; positive reactions were also registered on farms without case histories of BD. During consecutive observation and testing of seropositive horses, the infection became manifest in some animals; the majority, however, did not show clinical signs during the observation period of 5 years (Herzog et al. 1994). Clinically healthy but seropositive horses have been detected in Switzerland, the Netherlands, Poland, Luxemburg, Russia, Israel, and the United States (Kao et al. 1993; Herzog et al. 1994). In view of these findings, the assumption seems to be justified that natural infections by BDV remain subclinical in a relatively large number of cases and that endogenous or exogenous factors influence the outcome of the infection.

Assays for BDV-specific antibodies are conventionally carried out by immunofluorescence with persistently infected MDCK cells and by immunoblotting. Antibody titers in subclinically infected horses ranged between 1:5 and 1:500, analogous to titers in naturally infected horses with overt disease. In the limited number of samples of CSF tested so far, comparable antibody titers were found. Infectious virus could not be isolated from clinically normal seropositive horses. In a few preliminary studies, virus-specific RNA was detected by PCR in lachrymal fluid and/or nasal secretions (Richt et al. 1993; Herzog et al. 1994). This means that such animals should be regarded as virus carriers and potential sources of infection.

5 Experimental Transmission of BDV

A wide variety of animal species ranging from chickens to nonhuman primates can be infected experimentally with BDV. Only i.c. or i.n. inoculations result in predictable infections. Incubation periods and signs and severity of the disease depend on the animal species and on the virus variants used for infection. The pathohistological picture after experimental infections largely resembles that observed in natural cases. Even if great care is taken that the inoculum is deposited intracerebrally, some animals do not develop disease. Interestingly, most of these animals exhibit the typical morphological changes in the CNS and harbor infectious virus in their brain tissue.

The clinical picture in experimentally infected animals does not differ from that of spontaneously infected natural hosts. *Sheep* and *goats* seem to be less susceptible to i.c. inoculations than the horse. Experimental infection of *cattle* was not successful (Zwick 1929), but the disease could be provoked by a subsequent injection of a pyrogenic vaccine of *Escherichia coli* (Matthias 1958). In general terms, this observation suggests that stress factors lead to overt disease in asymptomatic infected animals. The recently described natural infections in cattle indicate that host-adaptive variations can occur.

Rabbits have been used as experimental hosts since the earliest experiments were carried out by Zwick and Seifried (1925). After an incubation period of 3–4 weeks most animals develop signs of depression, general fatigue and inattention; as the disease progresses, somnolence dominates the clinical picture. Paralysis of the legs follows, resulting in a decubitus posture. Rabbits become anorectic, emaciated, weak and ultimately lapse into coma and death. Degeneration of rods and cones and a gradual disappearance of neurons from the inner and outer nuclear layers of the retina causes blindness of the animals (Krey et al. 1979). The acute phase of the disease lasts 8–14 days.

Guinea pigs may have disease after i.c. or i.n. infections; however, they are less susceptible than rabbits (Zwick and Seifried 1925). The incubation period is considerably longer than for rabbits and may range from 19 to 373 days. The disease also progresses more slowly in guinea pigs. Clinical signs and the pathohistology generally correspond to the findings in naturally infected hosts. Antibodies can be demonstrated in sera from diseased or asymptomatic infected guinea pigs (von Sprockhoff 1957).

Chickens are frequently susceptible to BD (Zwick et al. 1927). In one of our own experimental series, the disease became manifest in about two thirds of the i.c. infected chickens. After an incubation period of about 18 months their movements became uncoordinated and stumbling, followed by paralysis and emaciation. Histological findings were BD-like cerebral lesions. Virus could be isolated in a few cases without clinical manifestations (unpublished results). Virulence of BDV did not increase after serial passages in the choriallantoic membrane of embryonated chicken eggs which is permissive for BDV replication (Rott and Nitzschke 1958).

Monkeys (*Macaca mulatta*) can fall ill after i.c., i.n. subdural or repeated subcutaneous inoculations after an incubation period of 4–8 weeks (Zwick et al. 1928). Symptoms are similar to those in rabbits: loss of attention, anorexia, occasionally somnolence, ataxia, convulsions and paralysis. In spite of infiltrations of mononuclear cells in the retina, similar to those in rabbits, no destruction of neuronal layers was observed (Stitz et al. 1980).

Tree shrews (*Tupaia glis*), classified phylogenetically at the root of primates, were found to develop a long-lasting, persistent productive infection in the CNS, resulting in an unusual, nonfatal behavioral disease after i.c. inoculation of BDV (Sprankel et al. 1978). Characteristic perivascular infiltration of mononuclear cells and intranuclear Joest-Degen inclusion bodies could be observed in the CNS. Housing conditions influenced the reaction of these animals after infection. The behavior of all paired animals changed dramatically. After the first few weeks following infection, these animals had the tendency to accept their partners much more quickly in grooming social interactions than did uninfected tree shrews. Normally passive females became as aggressive as their male partners and neglected their offspring. By contrast, only 25% of the infected animals that were kept in solitary cages showed clinical or behavioral changes. These animals exhibited hyperactivity with drastic shortening of their resting periods and developed distortions in their eating habits (bulimia) about 4 weeks after infection. The hyperactive phase was followed by a hypoactive period characterized by retiring into sleeping boxes and an increased tendency for self-grooming. Only a few of these animals developed explicit neurological signs such as ataxia and partial paralysis of the limbs (Sprankel et al. 1978).

5.1 Experimental Infection of Rats

The result of a BDV infection in the rat depends on the animal strain used and is influenced by the virus strain inoculated. Black-hooded (BH) rats do not develop any clinical signs after i.c. inoculation of some virus strains, in spite of virus persistence and infiltrations of mononuclear cells in the CNS. Crosses between a susceptible and a resistant rat strain (Lewis x BH rats) resulted in offspring that carried BDV resistance as a dominant trait and which was not associated with an IR or MHC gene. Genes regulating the differentiation of lymphocytes were not found to be responsible for susceptibility to BDV (Herzog et al. 1991).

Wistar rats do not develop clinical manifestations regularly after i.c. or i.n. inoculations. The highly variable incubation period, which can last for more than a year and is influenced by the age of the animal, can be shortened by serial passages in rat brain (Nitzschke 1963). Fatally infected animals exhibit typical paralysis of the extremities and the histopathological lesions seen in rabbits. Those rats, which remain healthy after infection, can harbor BDV permanently in their CNS and can produce virus-specific antibodies.

Lewis rats are highly susceptible to infection by BDV. As a result, this strain has been used extensively for studies in BD pathogenesis (Narayan et al. 1983). After i.c. injection, infectivity is first detected in brain homogenates 7 days after inoculation. The level of infectious virus reaches 10^5–10^6 ID_{50}/g brain by day 15 and this titer usually persists throughout the lifespan of the animal. After infection with an isolate from a horse, paralysis, obesity and/or fertility problems may occur. Distinct changes in the behavioral habits are the initial signs of disease in these infected animals. Hyperactivity lasts approximately 3 weeks, coincident with massive perivascular infiltrations in the limbic system. This phase remits to be replaced by apathy and depression. The initiation of this second phase is correlated with a stepwise decrease of inflammatory lesions in the CNS and the appearance of internal hydrocephalus. About 75 days postinfection, the hydrocephalus stabilizes and inflammatory lesions recede in spite of continued clinical disease and consistent virus titers. This self-limiting process, developing late after infection, has not been described in any other virus infection. In the eye, a retinitis develops that results in destruction of the pigment layer, photoreceptors and neural cells. The disappearance of these permissive host cells probably explains why infectious virus can no longer be demonstrated in the eye during late periods after infection.

5.2 Variability of BDV

Borna disease virus seems to have a high capacity for variability which may allow it to respond to the influence of host factors (Herzog et al., unpublished results). An equine isolate, which caused behavioral irregularities, underwent several passages in rabbits and three consecutive passages in rats. After further passages in newborn rats, the virus disappeared gradually, starting from day 5 postinfection, until it could no longer be demonstrated in brain tissue at 100 days postinfection by infectivity assays or immunohistological methods. Behavioral disorders were regularly replaced by obesity starting around day 30 postinfection and continued for an observation period of 2 years. In vitro passages of the virus in cerebral or lung cells derived from rat embryos also primes the virus for causing this type of obesity. However, when the virus was inoculated again into newborn rats, paralysis occurred at the age of 4 weeks and the majority of the animals died. Virus derived from persistently infected MDCK cells replicates in the CNS of the rat after i.c. inoculation, induces the production of virus-specific antibodies, but does not cause any clinical signs. One can conclude from these observations that the virus is capable of tolerating a series of mutations and that genetically stable variants can be selected from a heterogeneous virus population by host-derived factors. These new variants acquire different pathogenic properties.

The age of Lewis rats at the time of infection was found to have a decisive influence on the spread of the virus in the organism and the capacity of different cells to support virus replication. After infection of adult animals, infectious virus or viral antigen was exclusively found in neural tissue. In newborn animals, by contrast, the virus spread through the whole organism and could be demonstrated in most cells of peripheral organs without apparent disease (Narayan et al. 1983; Herzog et al. 1984). The reason for this variation in permissiveness is still unclear. Differences in the status of the animal's immune competence cannot be the only factor, since in adult immunosuppressed euthymic and athymic rats, the virus is strictly neurotropic (Herzog et al. 1985). These findings are of epidemiological interest, since neonatally infected animals are prone to excrete virus, in contrast to adult animals in which virus replication is restricted to neural tissue (Morales et al. 1988).

The observation by Krey et al. (1979), that in rabbits spread of the virus from the brain to the retina could be abrogated by ablation of the optic disk, furnished the initial concept regarding dissemination of BDV via neural pathways to and from the CNS. Incubation periods in rats varied greatly with the route of infection. The onset of BD only became manifest after the virus had invaded the brain. Clinical signs appeared 17 days after i.c. inoculation, 20–24 days after i.n. inoculation and 47 days after food-borne infection (Carbone et al. 1987).

5.3 Infection of Immunodeficient Rats

Infection of immunologically incompetent newborn or athymic rats, or animals which had undergone an immunosuppressive treatment by cyclophosphamide or cyclosporine A, did not result in signs of disease or inflammatory lesions in the CNS or the retina (Narayan et al. 1983; Stitz et al. 1989; Herzog et al. 1985) in spite of virus titers similar to those found in immunocompetent rats after an i.c. infection. Adoptive transfer of immunoreactive cells, particularly $CD4^+$ T cells, into immunodepressed infected rats caused the appearance of BD (Narayan et al. 1983; Richt et al. 1989; Stitz et al. 1989). The results of this series of experiments demonstrate that virus replication per se does not abrogate vital functions; rather the genesis of BD must be mediated by a virus-induced immunopathological reaction (see Stitz et al., this volume).

6 Nonsusceptible Animals

The following animal species did not develop clinical manifestation after BDV infections: ferrets (Nicolau and Galloway 1928), Syrian hamsters (Mattias 1955), pigeons (Zwick et al. 1927), and dogs (Zwick et al. 1927). However, infectious virus could be isolated from the CNS of apparently healthy infected Syrian hamsters (Anzil et al. 1978) and from the brain of a dog killed 158 days postinfection (Nicolau and Galloway 1930).

BDV replicates in the CNS of different strains of mice and can persist there for several months. In the cerebral tissue of some of these mice a moderate inflammation is apparent. Virus-specific antibodies are produced, but explicit disorders of the CNS do not occur (Heinig 1969; own unpublished results). Infected mice, however, may suffer from a deficit in learning capacities (Rubin et al. 1992) similar to disabilities observed in BDV-infected newborn rats (Dittrich et al. 1989). Learning disabilities may be related to a massive cellular degeneration in the stratum granulosum of the fascia dentata in the hippocampus.

7 Conclusions

A biphasic course for BD can be assumed in natural and experimental infections. After an unpredictable incubation period, the onset of a hyperactive phase is regularly observed, which can lead to a rapid death in some animals, particularly horses and rabbits. In animals which overcome this initial phase, a chronic hypoactive phase follows in conjunction with a regressing encephalitis and high levels of virus in the CNS. Signs of disease in the later phase may be so minimal as to suggest a full recovery. During this chronic phase, symptoms resembling those of the initial phase may suddenly reemerge. The obesity syndrome in rats is particularly remarkable, since there is a stepwise loss of infectious virus, viral antigen and virus-specific antibodies.

BDV seems to be variable in its pathogenic properties and in its capacity to form variants which can be selected during consecutive passages in different hosts giving rise to variable forms of the disease. The variable reactivities between monoclonal antibodies and viral antigens of different isolates leave the suspicion that a considerable number of infections may have passed unnoticed. The broad spectrum of species which can be infected justifies the assumption that BDV is harbored by more species than those currently known to be natural hosts.

In spite of i.c. inoculation of the virus, not all infected animals develop BD regularly, not even the most susceptible rabbits and Lewis rats. In some animal species, such as mice, hamsters, and BH rats, the infection regularly runs a subclinical or subtle course. BDV replicates in these inapparently infected individuals, persists in the CNS during prolonged periods, may cause pathological alterations in spite of the absence of clinical signs and stimulates the production of virus-specific antibodies. Sero-epidemiological studies suggest that, even in horses and sheep, natural infection by BDV remains clinically inapparent in a majority of cases.

The course of BD and the fate of the infected animals are dependent on factors such as the infectious dose and genetic properties of the virus and the age, immune status and genetic background of the host. These features have implications when considering the significance of virus-specific antibodies in humans. It is highly conceivable that humans are among the host species susceptible to infection by BDV and that this may result in no apparent disease, subtle abnormalities or profound disturbances in CNS function.

References

Altmann D, Kronberger H, Schüppel K-F, Lippmann R, Altmann J (eds) (1976) Bornasche Krankheit (Meningo-Encephalomeningitis Simplex Enzootica Equorum) bei Neuwelttylopoden und Equiden. Erkrankungen der Zootiere. Akademie, Berlin

Anzil AP, Blinzinger K, Mayr A (1973) Persistent Borna virus infection in adult hamsters. Arch Gesamte Virusforsch 40: 52–57

Beck H, Frohboese H (1926) Die enzootische Encephalitis des Schafes. Vergleichende experimentelle Untersuchungen über die seuchenhafte Gehirnrückenmarksentzündung der Pferde und Schafe. Arch Wiss Prakt Tierheilkd 54: 84–100

Caplazi P, Waldvogel A, Stitz L, Braun N, Ehrensperger F (1994) Borna disease in naturally infected cattle. J Comp Path 111: 65–72

Carbone KM, Duchala CS, Griffin JW, Kincaid AL, Narayan O (1987) Pathogenesis of Borna disease in rats: evidence that intra-axonal spread is the major route for virus dissemination and the determination for disease incubation. J Virol 61: 3431–3440

Carbone, KM, Moench TR, Lipkin WI (1991) Borna disease virus replicates in astrocytes, Schwann cells and ependymal cells in persistently infected rats: location of viral genomic and messenger RNAs by in situ hybridization. J Neuropathol Exp Neurol 50: 205–214

Daubney R, Mahlau EA (1967) Viral encephalomyelitits of equines and domestic ruminants in the Near East: part 1. Res Vet Sci 8: 375–397

Dittrich W, Bode L, Ludwig H, Kao M, Schneider K (1989) Learning deficiencies in Borna disease virus-infected but clinically healthy rats. Biol Psychiatry 26: 818–828

Elford WJ, Galloway IA (1933) Filtration of the virus of Borna disease through graded collodion membranes. Br J Exp Pathol 14: 196

Ernst W, Hahn W (1927) Weitere Beiträge zur Bornaschen Krankheit der Pferde und zur Frage der Ätiologie des bösartigen Katarrhalfiebers der Rinder. Munch Tierarztl Wochenschr 78: 85–89

Gosztonyi G, Ludwig H (1984) Borna disease of horses: an immuno-histological and virological study of naturally infected anmals. Acta Neuropathol (Berl) 64: 213–221

Haas B, Becht H, Rott R (1986) Purifcation and properties of an intranuclear virus-specific antigen from tissue infected with Borna disease virus. J Gen Virol 67: 235–241

Heinig A (1969) Die Bornasche Krankheit der Pferde und Schafe. In: Röhrer H (ed) Handbuch der Virusinfektionen bei Tieren, vol 4. Fischer, Jena, pp 83–148

Herzog S, Rott R (1980) Replication of Borna disease virus in cell culture. Med Microbiol Immunol 168: 153–158

Herzog S, Kompter C, Frese K, Rott R (1984) Replication of Borna disease virus in rats: age-dependent differences in tissue distribution. Med Microbiol Immunol 173: 171–177

Herzog S, Wonigeit K, Frese K, Hedrich HJ, Rott R (1985) Effect of Borna disease virus infection on athymic rats. J Gen Virol 66: 503–508

Herzog S, Frese K, Rott R (1991) Studies on the genetic control of resistance of Black Hooded rats to Borna disease. J Gen Virol 72: 535–540

Herzog S, Frese K, Richt JA, Rott R (1994) Ein Beitrag zur Epizootologie der Bornaschen Krankheit beim Pferd. Wien tierärztl Wschr (in press)

Hiepe T (1958/59) Die Bornasche Krankheit. Klinisch-diagnostische Untersuchungen an Pferden und Schafen mit besonderer Berücksichtigung des Liquor cerebrospinalis. Wiss Z Univ Leipzi 8: 263–338

Hirano N, Kao M, Ludwig H (1983) Persistent tolerant or subacute infection in Borna disease virus infected rats J Gen Virol 64: 1521–1530

Joest E, Degen H (1911) Untersuchungen über die pathologische Histologie, Pathogenese und postmortale Diagnose der seuchenhaften Gehirn-Rückenmarksentzündung (Bornasche Krankheit) des Pferdes. Z Inf Krkh Haustiere 9: 1–98

Kao M, Ludwig H, Gosztonyi G (1984) Adaptation of Borna disease virus to the mouse. J Gen Virol 65: 1845–1849

Kao M, Hamir HN, Rupprecht CE, Fu ZF, Shankar K, Koprowski H, Dietzschold B (1993) Detection of antibodies to Borna disease virus in sera and cerebrospinal fluid of horses in the USA. Vet Rec 132: 241–244

Krey HF, Ludwig H, Rott R (1979) Spread of infectious virus along the optic nerve into the retina in Borna disease virus-infected rabbits. Arch Virol 61: 283–288

Lange H, Herzog S, Herbst W, Schliesser T (1987) Seroepidemiologische Untersuchungen zur Bornaschen Krankheit (Ansteckende Gehirn-Rücken-markentzündung) der Pferde. Tierarztl Umschau 42: 938–946

Ludwig H, Becht H, Groh L (1973) Borna disease (BD) a slow virus infection. Biological properties of the virus. Med Microbiol Immunol 158: 275–289

Ludwig H, Bode L, Gosztonyi G (1988) Borna disease: a persistent virus infection of the central nervous system. Prog Med Virol 35: 107–151

Lundgren AL, Czech G, Bode L, Ludwig H (1993) Natural Borna disease in domestic animlas others than horses and sheep. J Vet Med B40: 298–303

Malkinson M, Weisman Y, Ashash E, Bode L, Ludwig H (1993) Borna disease in ostriches. Vet Rec 133: 304

Matthias D (1955) Neue Forschungsergebnisse bei der Bornaschen Krankheit der Pferde. Monatsh Vet Med 10: 123–126

Matthias D (1958) Weitere Untersuchungen zur Bornaschen Krankheit der Pferde und Schafe. Arch Exp Vet Med 15: 643–646

Mayr A, Danner K (1972) Production of Borna virus in tissue culture. Proc Soc Exp Biol Med 140: 511–515

Metzler A, Ehrensperger F, Wyler R (1978) Natürliche Bornavirus-Infektion beim Kaninchen. Zentralbl Vet Med B 25: 161–164

Morales JA, Herzog S, Kompter C, Frese K, Rott R (1988) Axonal transport of Borna disease virus along the olfactory pathways in spontaneously and experimentally infected rats. Med Microbiol Immunol 177: 51–68

Narayan O, Herzog S, Frese K, Scheefers H, Rott R (1983) Pathogenesis of Borna disease in rats: immune-mediated viral ophthalmoencephalopathy causing blindness and behavioral abnormalities. J Infect Dis 148: 305–315

Nicolau S, Galloway IA (1928) Borna disease and enzootic encephalomyelitis of sheep and cattle. Spec Rep Ser Med Res (Counc) 121: 7–90

Nicolau S, Galloway IA (1930) L'encéphalo-myélite enzootique expérimentale (maladie de Borna). Ann Inst Pasteur 44: 473-496

Nitzschke E (1963) Untersuchungen über die experimentelle Bornavirus-Infektion bei der Ratte. Zentralbl Vet Med B 10: 470–527

Otta J, Jentsch KD (1960) Spontane Infektion mit dem Virus der Bornaschen Krankheit bei Kaninchen. Monatsh Vet Med 15: 127–129

Richt JA, Stitz L, Wekerle H, Rott R (1989) Borna disease, a progressive meningoencephalomyelitis as a model for CDH⁺ cell-mediated immunopathology in the brain. J Exp Med 170: 1045–1050

Richt JA, VandeWoude S, Zink MC, Clements JE, Herzog S, Stitz L, Rott R, Narayan O (1992) Infection with Borna disease virus: molecular and immunobiological characterization of the agent. Clin Infect Dis 14: 1240–1250

Richt JA, Herzog S, Haberzettl K, Rott R (1993) Demonstration of Borna disease virus-specific RNA in secretions of naturally infected horses by the polymerase chain reaction. 182: 293–304

Rott R, Nitzschke E (1958) Untersuchungen über die Züchtung des Virus der Bornaschen Krankheit im bebrüteten Hühnerei unter verchiedenen Bedingungen. Zentralbl Vet Med 5: 629–633

Rott R, Herzog S, Bechter K, Frese K (1991) Borna disease, a possible hazard for man? Arch Virol 118: 143–149

Rubin SA, Waltrip RW, Bautista JR, Carbone KM (1992) Borna disease in mice: host-specific differences in disease expression. J Virol 67: 548–552

Schädler R, Diringer H, Ludwig H (1985) Isolation and characterization of a 14500 molecular weight protein from brains and tissue cultures persistently infected with Borna disease virus. J Gen Virol 66: 2479–2484

Schulze W (ed) (1951) Leitfaden der Ziegenkrankheiten. Hirzel, Leipzig.

Seifried O, Spatz H (1930) Die Ausbreitung der encephalitischen Reaktion bei der Bornaschen Krankheit der Pferde und deren Beziehung zu der Encephalitis epidemica, der Heine-Medinschen Krankheit und der Lyssa des Menschen. Eine vergleichend-pathologische Studie. Z Neurol Psychiatry 124: 317–382

Sprankel H, Richarz K, Ludwig H, Rott R (1978) Behavior abnormalities in tree shrews (tupaia glis, Diard, 1920) induced by Borna disease virus. Med Microbiol Immunol 165: 1–18

Stitz L, Krey H, Ludwig H (1980) Borna disease in rhesus monkeys as a model for uveo-cerebral symptoms. J Med Virol 6: 333–340

Stitz L, Soeder D, Deschl U, Frese K, Rott R (1989) Inhibition of immune-mediated meningoencephalitis in persistently infected Borna disease virus-infected rats by Cyclosporin A. J Immunol 143: 4250–4256

Stitz L, Schilken D, Frese K (1991) Atypical dissemination of the highly neurotropic Borna disease virus during persistent infection in Cyclosporin A-treated, immunosuppressed rats. J Virol 65: 457–460

Thiedemann N, Presek P, Rott R, Stitz L (1992) Antigenic relationship and further characterization of two major Borna disease virus proteins. J Gen Virol 73: 1057–1064

Von Rheinbaben F, Stitz L, Rott R (1985) Influence of interferon on persistent infection caused by Borna disease virus in vitro. J Gen Virol 66: 2777–2780

Von Sprockhoff H (1957) Über das Vorkommen der Joes-Degenschen Einschlußkörper bei Bornavirus-infizierten Kaninchen, Meerschweinchen und Ratten. Alters. Monatsh Prakt Tierhkd 8: 129–142

Zimmermann W, Breter H, Rudolph M, Ludwig H (1994) Borna disease virus: immunoelectron microscopic characterization of cell-free virus and further information about the genome. J Virol 68: 6755–6758

Zwick W (1939) Bornasche Krankheit und Encephalomyelitis der Tiere. In: Gildemeister E, Haagen E, Waldmann O (eds) Handbuch der Viruskranskheiten, vol 2. Gustav Fischer, Jena, pp 254–356

Zwick W, Seifried O (1925) Übertragbarkeit der seuchenhaften Gehirnrückenmarksentzündung des Pferdes (Bornasche Krankheit) auf kleine Versuchstiere (Kaninchen). Berl Tierarztl Wochenschr 41: 129–132

Zwick W, Seifried O, Witte J (1927) Experimentelle Untersuchungen über die seuchenhafte Gehirn- und Rückenmarksentzündung der Pferde (Bornasche Krankheit) Z Inf Krkh Haustiere 30: 42–136

Zwick W, Seifried O, Witte J (1928) Weitere Untersuchungen über die Gehirn-Rückenmarksentzündung der Pferde (Bornasche Krankheit). Z Inf Krkh Haustiere 32: 150–179

A Borna-Like Disease of Ostriches in Israel*

M. Malkinson[1], Y. Weisman[1], S. Perl[1], and E. Ashash[2]

1 Introduction . . . 31
2 Clinical Disease . . . 32
3 Pathology . . . 32
4 Pathogenesis . . . 33
5 Virus Isolation . . . 33
6 Serology . . . 35
7 Infection Studies . . . 35
8 Preventive Therapy . . . 36
9 Electron Microscopy . . . 36
10 Discussion . . . 37
References . . . 38

1 Introduction

Wild ostriches were once part of the indigenous fauna of the Middle East. The earliest archaeological artifacts date from the Chalcolithic era (4000 B.C.E.) or earlier (Grigson 1987; Paz 1987), when *Struthio camelus syriacus*, an extinct subspecies, inhabited the Fertile Crescent. All the other extant subspecies are native to Africa. The ostrich is mentioned several times in the Bible as an inhabitant of the desert. In Israel, ostrich breeding farms are a recent addition to the agricultural scene. Young birds were imported from Africa in the early 1980s for breeding purposes and by 1987 six breeding farms had been established. There is now an estimated total of 10 000 adult breeding birds in Israel.

[1] Kimron Veterinary Institute, Beit Dagan, POB 12, Israel
[2] Ashdot-Yaacov, Jordan Valley, Israel

*Some of the information reported here has been published as "Letters to the Editor," in Veterinary Record (1993) 131: 284; 132: 172, 304.

Fig. 1. Paretic ostrich attempting to stand

2 Clinical Disease

The paretic condition that is now attributed to Borna disease virus first appeared in young ostriches on several farms in 1988. Birds between 14 and 42 days old were mostly affected. In a minority of cases, clinical signs of incoordination were seen 1–3 days before paresis supervened. At this stage of the disease the birds could stand and move with difficulty if given external support (Fig. 1). Mostly there was no premonition of the disease. Appetite, vision and hearing of the affected birds were normal but secondary infections and infected pressure sores resulted in their demise within 1–3 weeks of sternal recumbency. About 1% of the ostriches seemed to recover if they were given intensive supportive care; however, they relapsed several months later.

Between 1989 and 1992, when detailed records were kept, 7%–16% of all the chicks that hatched died from paresis. On one of the worst affected farms, for example, out of a total of 1331 ostriches received, 486 died by the age of 3 months; of these, 181 were paretic. The syndrome was therefore the most significant cause of morbidity in young ostriches during this 4 year period. Thereafter, with the introduction of serum therapy, the situation improved considerably.

3 Pathology

No specific lesions were visible on gross necropsy other than that the cloaca was filled with a voluminous quantity of greenish-yellow liquid feces. In all the field cases, microscopic lesions were confined to the lumbosacral region of the spinal cord where the affected cell population consisted of neurons located primarily in the central gray matter (Fig. 2). The lesions ranged from degenerative changes accompanied by numerous infiltrating glial cells (satellitosis) and neuronophagia to the formation of glial nodules that replaced the neuron remnants. In experimentally induced cases, perivascular cuffing was a notable feature.

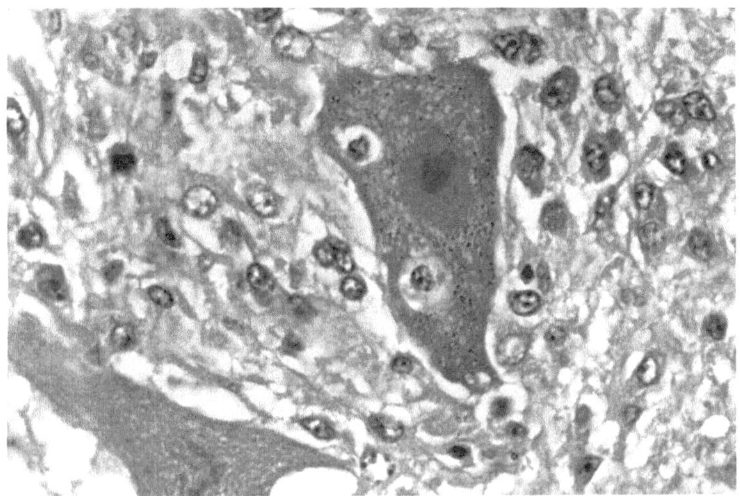

Fig. 2. Motor neuron at an early stage of degeneration with two glial cells penetrating the cytoplasm

4 Pathogenesis

All attempts to identify a noninfectious cause yielded inconclusive results. These included biochemical, nutritional, genetic and ecological or toxicological lines of investigation. Infectious agents such as the commonly associated causes of nervous diseases of birds were also negative. These included *Pasteurella* spp., *Chlamydia*, avian encephalomyelitis virus, Newcastle disease virus, turkey meningoencephalitis virus and other arboviruses (eastern and western equine encephalomyelitis). Other less commonly encountered neurotropic viruses were therefore considered. Of special geographical interest was Borna disease virus (BDV). This agent had been isolated from the brains of wild birds caught in the Syrian countryside in the neighborhood of farm animals suffering from outbreaks of a nervous disease (Daubney and Mahlau 1967). Brains were therefore harvested from sick and healthy ostriches and sent to Professor Hanns Ludwig, Head of the Borna Disease Virus Reference Laboratory, Robert Koch Institute, Berlin. BDV antigen was detected in seven of the 13 brains from sick ostriches and in only one brain from 10 healthy ostriches that were examined by ELISA.

5 Virus Isolation

Spinal cord, brain and spleen of a freshly killed paretic ostrich were trypsinized and cocultivated with MA104 cells (monkey kidney cell line). Following several blind passages in which no cytopathic effect was detected, monolayers were fixed with cold acetone and stained by immunofluorescence. Bizarre patches of fluorescence were observed (Fig. 3). Further passages were made with the C6 rat astrocytoma cell line. In these cells, fluorescent granules were visible in the nuclei (Fig. 4). Both these preparations were stained with an adult ostrich serum originating from an infected farm.

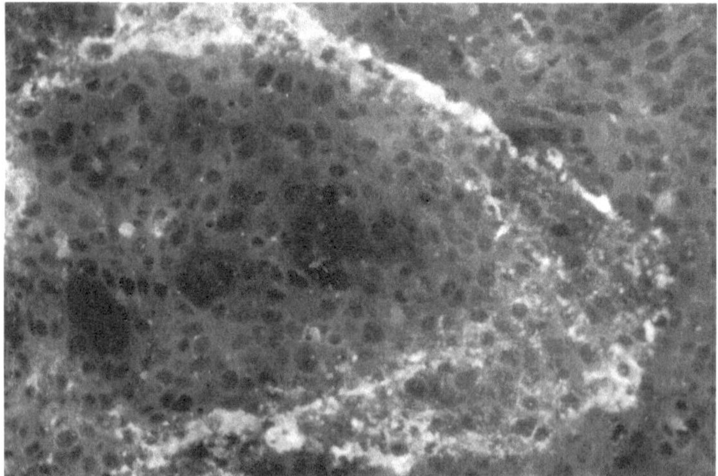

Fig. 3. Immunofluorescence of MA104 cells at the fourth blind passage of a spleen cocultivate derived from a paretic ostrich

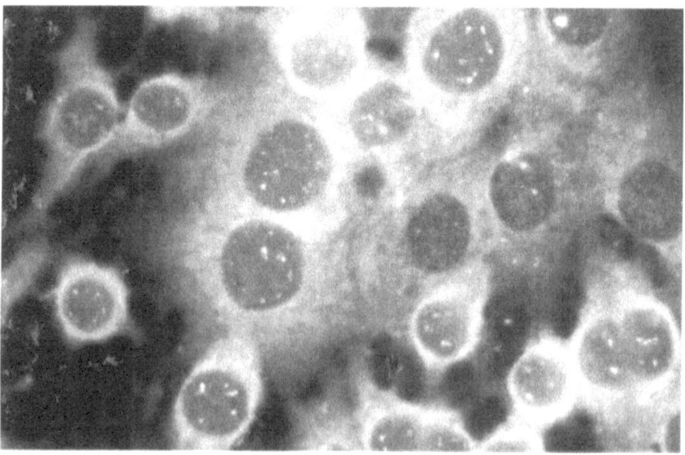

Fig. 4. Immunofluorescence of C6 cells at the second blind passage of buffy coat cells from paretic ostrich blood

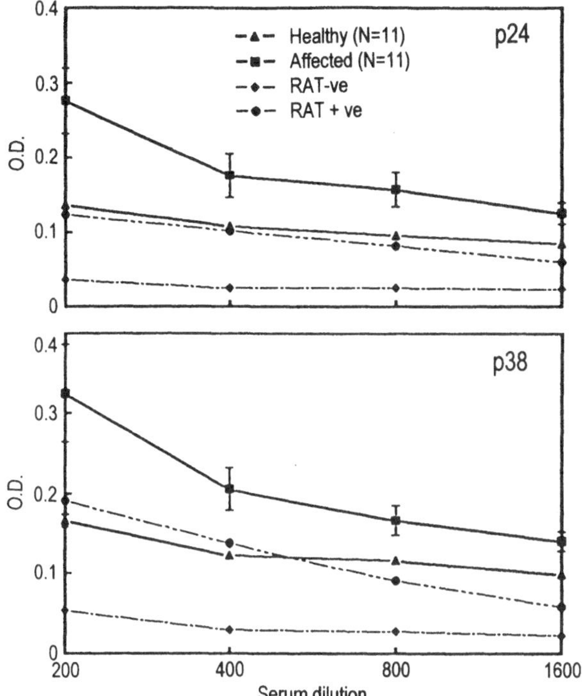

Fig. 5. ELISA study of two panels of sera from ostriches raised on farms with and without a clinical history of paresis using plates coated with recombinant p24 and p38 BDV proteins. Sera from BDV-infected and noninfected rats were included as controls

6 Serology

Based on the detection of BDV antigen in infected brain, we have attempted to develop an ELISA for ostrich antibodies by extracting antigen from the brains of BDV-infected ostriches and sheep (MALKINSON et al. 1994). Initial results indicate that sheep brain contains more antigen than does ostrich brain and that both antigens react similarly with panels of positive and negative ostrich sera. These panels were also tested by ELISA using plates coated with recombinant p24 and p38 BDV proteins (supplied by Professor Ian Lipkin). Figure 5 shows that both proteins could differentiate between the positive and negative panels.

7 Infection Studies

A cohort of 21,5-week-old ostriches was divided into three groups; one group of five birds was injected intramuscularly and a second group of six birds was infected orally with a brain homogenate prepared from two paretic ostriches that were BDV antigen-positive. A third group of ten birds served as uninfected controls. Clinical signs of paresis indistinguishable from the field cases, i.e., sudden onset, unimpaired vision and appetite, appeared in four of the intramuscularly infected group and in four of the orally infected birds. The birds were euthanized in extremis within 3 days of the onset of disease. The mean death times for each subgroup were 17 and 42 days, respectively. No clinical disease occured in the uninfected controls through a 90 day period of observation.

8 Preventive Therapy

Pooled serum was prepared by exsanguinating ostriches 7–10 days from the onset of paresis. A similar pool was prepared from 11-month-old clinically normal adults. In a controlled experiment on one ostrich farm with a history of high morbidity, four successive batches of 30 chicks were injected at 2, 7, 14 and 21 days of age with 0.5, 1.0, 1.5 and 2.0 ml, respectively, of serum. Half the number received paretic ostrich serum while the rest received the adult serum pool. A total of six and five birds died in each treatment group, with mean death times of 68.9 + 6.1 (standard error) and 25.6 + 7.5 days, respectively ($p < 0.001$; *t* test). Thus, the paretic ostrich pooled serum was capable of substantially delaying the onset of clinical signs but was not completely protective.

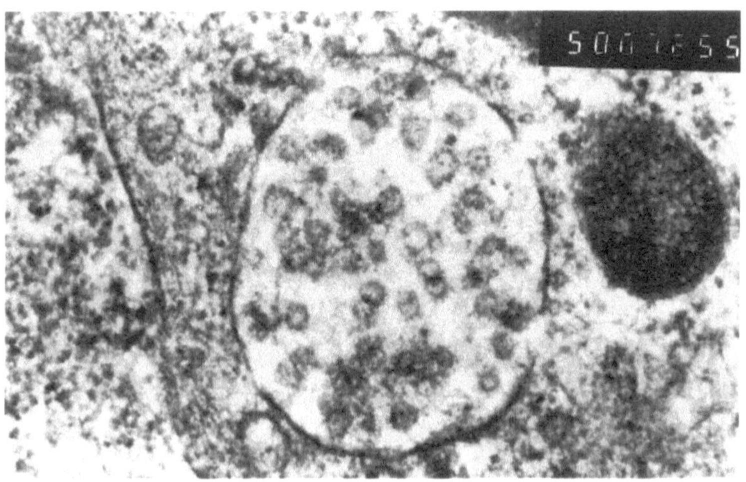

Fig. 6. Electron micrograph of infected MA104 cells (as in Fig. 3) showing a vacuole filled with spherical structures. x50 000

9 Electron Microscopy

Thin sections of the immunofluorescence-positive, MA104 cells revealed cytoplasmic vacuoles containing spherical viral particles. These appear to possess a bilayered envelope with poorly defined fringe-like projections. No particles were seen budding from the surface membrane (Fig. 6).

10 Discussion

Our research on ostrich paresis has now reached the point where we can be justified in attributing its cause to BDV or a closely related virus. Though the ostrich is the first natural avian host to be affected clinically by the virus, chickens have been experimentally infected (LUDWIG et al. 1988). In chickens, persistent infection resulted in ataxia over an extended period of time. The detection of BDV antigen in brains of paretic ostriches was also the first step in fixing a viral etiology for the disease, and this finding justified employing brain homogenate for infecting the young ostriches. From field cases we suspected that infection occurs via the oral-fecal route, while others factors, such as the presence of maternal immunity, govern the extended age incidence. The orally infected birds became sick at different times after exposure to the virus, which would indicate that the pathogenesis of the disease involves a replication site for the virus outside of the central nervous system. Further work is required to determine whether this site is the gastrointestinal tract or the nasopharynx. Epidemiological findings, based on a comparison of the incidence of disease among ostrich-rearing farms, indicate that the virus is not egg-transmitted.

Another feature of the virus isolated from the ostrich which is similar to BDV is its behavior in cell culture; cocultivation and blind passages were required for its isolation and absence of cytopathic effect. The location of fluorescence in both the nucleus and cytoplasm according to the cell line and possibly the time course of infection is also similar to BDV. Work in progress is directed at cross-immunofluorescence studies using a panel of BDV antibodies from the mouse, rabbit, rat and sheep. Infectivity experiments with tissue culture-passaged virus in other avian species are planned in order to ascertain whether the particles seen on electron microscopy are causative agents.

Preliminary results with the ELISA show some degree of cross-reactivity between sheep and ostrich brain antigens and between the purified p24 and p38 BDV proteins vis-a-vis the two panels of ostrich antisera. Further ELISA studies should indicate whether maternal antibodies play a role in the natural history of the disease.

The gross and histopathological findings in avian species are substantially different from the mammalian experience. For example, perivascular cuffing was observed only in the spinal cords of experimentally infected birds and no characteristic lesions have been seen in the brains of paretic ostriches. No Joest-Degen inclusion bodies have been identified so far. These differences may be phylogenetically linked and possibly related to the immune response of the ostrich, about which we have minimal information.

Finally, BDV in Israel appears not to be limited to the ostrich. There is accumulating evidence for the presence of the virus as a sporadic infection in livestock, including horses and sheep (Malkinson et al. 1994).

Acknowledgments. We wish to thank Professor Hanns Ludwig for his seminal contribution to this study and to Dr. Kathryn Carbone and Professor W. Ian Lipkin for providing diagnostic reagents. We are grateful to Professor Abraham Shahar for electron microscopy. Raya Blumenkranz and Rosi Meir gave unstinted technical assistance and Ilya Shkap provided photographic services.

References

Grigson C (1987) Shiqmim: pastoralism and other aspects of animal management in the Chalcolithic of the northern Negev. In: Levy TE (ed) Shiqmim I: Studies concerning Chalcolitic societies in the Northern Negev desert, Israel (1982–1984). British Archeological Trust, Oxford (BAR international series 356, Chap. 7, p 219)

Paz U (1987) The birds of Israel: order: struthioniformes (ostriches) Greene, Lexington, p 11

Daubney R, Mahlau EA (1967) Viral encephalomyelitis of equines and domestic ruminants in the Near East. Res Vet Sci 8: 375

Malkinson M, Rapoport E, Ludwig H (1994) Borna disease in sheep: first case recorded in Israel. Isr J Vet Med 49 (in press)

Ludwig H, Bode L, Gosztonyi G (1988) Borna disease: a persistent virus infection of the central nervous system. Prog Med Virol 35: 107

Borna Disease — Neuropathology and Pathogenesis

G. GOSZTONYI[1] and H. LUDWIG[2]

1 Introduction . . . 39

2 The Natural Disease . . . 40
2.1 Character of the Inflammatory Reaction in the Central Nervous System . . . 40
2.2 Neuronal Damage . . . 42
2.3 Demonstration of Virus-Specific Antigens and Nucleic Acids . . . 43
2.4 Electron Microscopy . . . 45
2.5 Regional Distribution of Histological Lesions and Viral Products . . . 45
2.6 Natural Infection of Sheep, Cats and Ostriches . . . 46

3 Experimental Infections . . . 47
3.1 Experimental Borna Disease of the Rat . . . 47
3.1.1 Types of Experimental Disease . . . 47
3.1.2 Persistent Infection . . . 48
3.1.3 Hyperacute Infection . . . 59
3.1.4 Acute/Subacute Infection . . . 60
3.1.5 Intracerebral, Intraocular and Peripheral Inoculations . . . 60
3.1.6 Obesity Syndrome . . . 61
3.2 Experimental Borna Disease of Mice . . . 62
3.3 Experimental Borna Disease in Rabbits, Hamsters, Tree Shrews, Monkeys and Chicken . . . 62

4 Pathogenesis . . . 63
4.1 Virus Spread . . . 63
4.2 Borna Disease Virus and Neurotransmitter Receptors . . . 63
4.3 Borna Disease Virus Antigens in the Hippocampal Formation: Their Affinity for Aspartate and Glutamate Receptors . . . 64
4.4 Cell and Tissue Tropism of Borna Disease Virus: Elective Vulnerability . . . 65
4.5 Phasic Expression of Viral Proteins . . . 66
4.6 Obesity Syndrome . . . 66
4.7 The Role of Cellular Immunity . . . 67

5 Summary . . . 68

References . . . 69

1 Introduction

Histopathology has contributed substantially to the understanding of the nature of Borna disease (SEIFRIED and SPATZ 1930; GOSZTONYI and LUDWIG 1984a). The Borna

[1] Institut für Neuropathologie, Freie Universität Berlin, Universitätsklinikum Benjamin Franklin, Hindenburgdamm 30, 12200 Berlin, Germany
[2] Institut für Virologie, Freie Universität Berlin, Nordufer 20 (Robert Koch-Institut), 13353 Berlin, Germany

disease virus (BDV) belongs to the single- and negative-stranded RNA viruses (LIPKIN et al. 1990; DE LA TORRE et al. 1990; BRIESE et al. 1992; SCHNEIDER et al. 1994); however, since it has not yet been visualized morphologically, its classification remains to be established. Although BDV has features of a conventional enveloped virus, it behaves in many respects in a way that has not been observed in any other viral agent. Its primary target is the nervous system, but it appears and apparently replicates in many other organs as well. Its spread is neural, i.e., axonal and transsynaptic. Depending on the virulence of the agent and on the immunological status of the host, several clinical courses of the disease and several types of histopathological manifestations are known (LUDWIG et al. 1985, 1988). In nature, the disease occurs in horses and sheep. A similar syndrome was recently found in cats (LUNDGREN and LUDWIG 1993; LUNDGREN et al. 1993) and ostriches (MALKINSON et al. 1993). Borna disease can be transmitted experimentally to a series of animal species from birds to primates (LUDWIG et. al.1988). Serological studies indicate that human BDV infections also exist (see chapter by Bode).

2 The Natural Disease

In its natural form, Borna disease (BD) is a nonpurulent encephalomyelitis with a predilection to involve the gray matter of the cerebral hemispheres and the brain stem (SEIFRIED and SPATZ 1930; GOSZTONYI and LUDWIG 1984a). Its histopathology became known by detailed studies on the central nervous system (CNS) of diseased horses (JOEST and DEGEN 1911). Inflammation is accompanied by a moderate degeneration of ganglion cells. With the exception of the hemorrhagic form of the disease, no significant macroscopic changes can be found in the brain and spinal cord. No pathological alterations are detectable in the visceral organs at general autopsy.

2.1 Character of the Inflammatory Reaction in the Central Nervous System

The inflammatory cells appear in the form of perivascular cuffs of various widths around venules and small veins, and, rather infrequently, around small arteries. They occupy, as a rule, the adventitial space of the vessels but occasionally break through the adventitial membrane to form perivascular infiltrates (Fig. 1A). The bulk of the infiltrating cells consists of lymphocytes, monocytes, and to a lesser extent, plasma cells. Polymorphonuclear leukocytes have rarely been seen (JOEST and DEGEN 1911; SEIFRIED and SPATZ 1930; SEIFRIED 1931).

The perivascular infiltrates may turn into more diffuse interstitial or tissue infiltrates. The dominant cell types in the latter are the monocytes and microglial cells. They are usually slender and transform only occasionally into plump

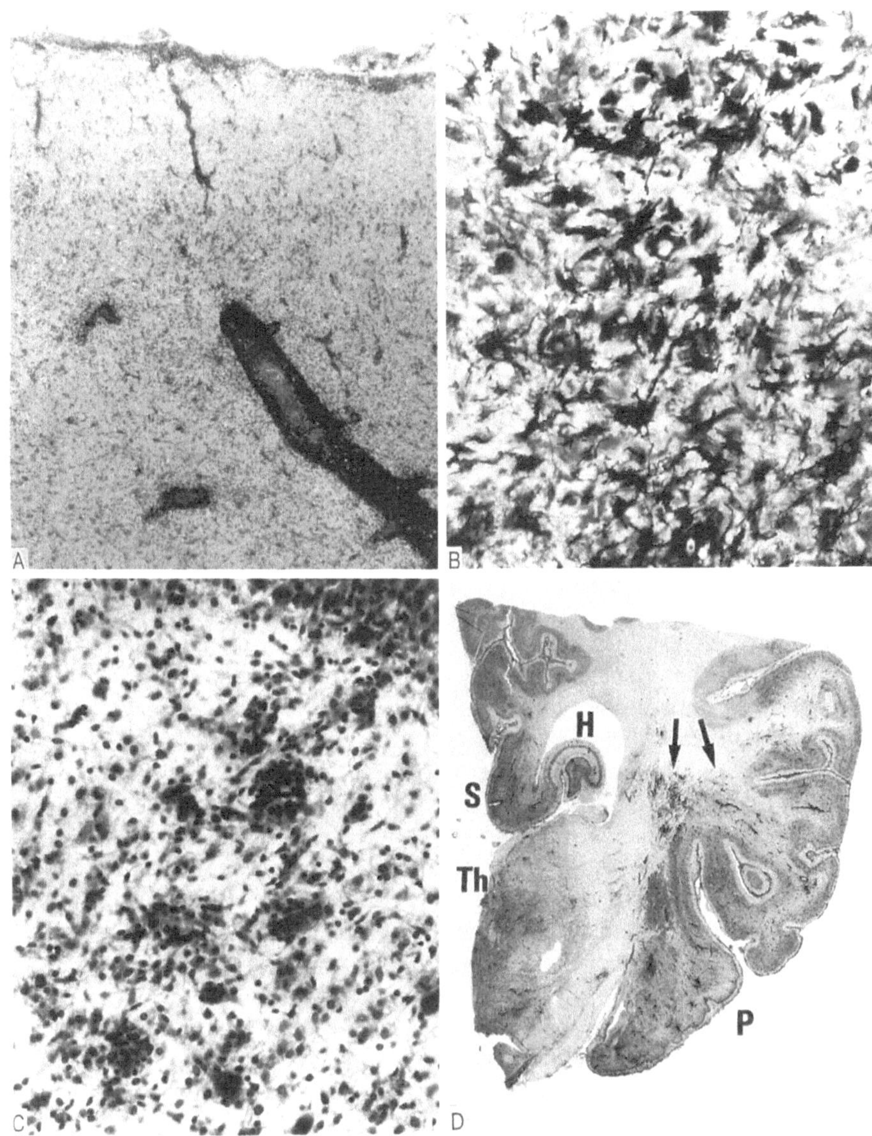

Fig. 1A–D. Borna encephalitis of the horse. **A** Adventitial and perivascular inflammatory infiltrates in the cerebral cortex. The massive infiltrate around a small cortical vein is particularly conspicuous. Only slight inflammatory infiltrates on the leptomeninges. H & E stain, x 46. **B** Diffuse tissue infiltrate: a cluster of slender and plump microglial cells in the hippocampus. Gallyas' silver impregnation technique for microglia, x 290. **C** Neuronophagic nodules around diseased hippocampal pyramidal cells. Nissl stain, x 290. **D** Coronal section through the right cerebral hemisphere of the horse brain. Severe inflammatory infiltrates in the hippocampus (*H*), subiculum (*S*), paraventricular parts of the thalamus (*Th*), pyriform cortex (*P*) and adjacent neocortical areas. The white matter under the severely involved cortical areas also contains inflammatory infiltrates (*arrows*). Celloidin technique, Nissl stain, x 1.9

microglia (Fig. 1B) and fat granule cells. Occasionally, microglial cells accumulate around diseased neurons, forming neuronophagic nodules (Fig. 1C). In BD, however, neuronophagias are less frequently encountered than in other primary viral encephalitides. In the white matter, small accumulations of microglia, "glial stars," are randomly found. The majority of the perivascular inflammatory infiltrates is found in the gray matter, but the underlying white matter also harbors perivascular cuffs of inflammatory cells (Fig. 1D).

A slight leptomeningitis regularly accompanies the encephalitis (Fig. 1D). This inflammatory reaction, and the presence of oligoclonal antibodies in the cerebrospinal fluid (CSF) (Ludwig and Thein 1977; Ludwig et al. 1993), contribute to the clinical diagnosis of the disease.

The acute phase of the illness is followed by a progressive hyperplasia of the astroglia, mainly in the gray matter. Both fibrous and protoplasmic astrocytes are seen among the hyperplastic cells. In the late phase of the disease a fibrillary gliosis can be found in the more severely affected areas.

2.2 Neuronal Damage

One of the peculiar features of BD is that only a slight and rare neuronal degeneration accompanies the heavy inflammatory reaction. Even nerve cells surrounded by inflammatory cells seem to be morphologically intact at light microscopic examination. An early accomplishment was the description of inclusion bodies in the nuclei of neurons by Joest and Degen (1909), since then regarded as pathognomonic of the disease (Fig. 2A). These authors used the Mann staining technique modified by Lentz (Bohne 1907). The intranuclear inclusions were bright red and surrounded by a halo. The inclusions may be solitary or multiple in the same nucleus, with a diameter varying between 1 and 6 µm. There is no change in the appearance of the nucleolus and the rest of the cell. These inclusions bodies occur most frequently in the large neurons, e.g., in the pyramidal cells of the hippocampus.

In the earlier literature, the existence of degenerative neuronal changes was controversial. Gross signs of neuronal damage could not reliably be assessed in equine postmortem brain material because of less elaborate fixation; however, studies in experimentally infected animals confirmed that neuronal damage occurs early in BD. In areas where inflammatory changes are severe, nutrition of neurons may suffer, probably because cellular infiltration of the vessel walls and inflammatory edema hinder the transport processes. Neuronal degeneration in such areas is quite obvious and can reach the severity of incomplete necrosis. In the horse retina, severe degeneration of the neurons has been demonstrated by several investigators. The development of amblyopia and amaurosis in horses was first described by Schmidt (1912) and later confirmed by Zwick (1939), Walther (1952) and Müller and Fritsch (1955). The first pathological description of the eye (Walther 1952) reported a nonpurulent, lymphocytic infiltration of the retina with damage to the neurons and degeneration of the optic nerve. The

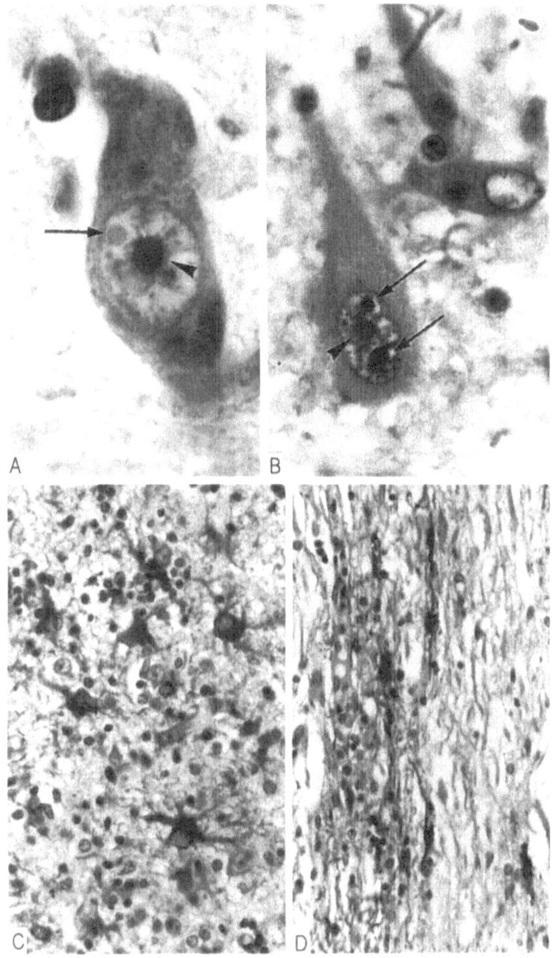

Fig. 2 A–D. Borna encephalitis of the horse. **A** Pyramidal cell of the hippocampus. Small intranuclear eosinophilic inclusion body of Joest-Degen with a clear halo (*thin arrow*) in the neighborhood of an intact nucleolus (*arrowhead*). H & E stain, x 1056. **B** Pyramidal cell of the hippocampus. Immunostaining with a polyclonal anti-BDV antibody. The cytoplasm of the neuron is free from viral antigen. The nucleus contains two round aggregates of virus-specific material (*slender arrows*), corresponding to the Joest-Degen inclusion bodies. The nucleolus (*arrowhead*) is distinctly negative. PAP technique, x 605. **C** Reactive astrocytes in the hippocampus. Their cytoplasm is diffusely positive for BDV-specific antigen. Immunostaining with a monoclonal anti-BDV antibody, avidin-biotin-peroxidase (ABC) technique, x 240. **D** Longitudinal section of a fiber tract in the mesencephalon. The cell bodies and processes of several slender interfascicular oligodendroglial cells are positive for BDV-specific antigen. x 240

latter findings were more pronounced in horses with longer survival. Titers of infectious virus as high as 2×10^4 ffu have been found in the retina of horses with BD.

2.3 Demonstration of Virus-Specific Antigens and Nucleic Acids

The presence and distribution of virus-specific antigen has been studied immunohistochemically in the horse brain using fluorescent antibody (Wagner et al. 1968) and unlabeled antibody-enzyme techniques with both poly- and monoclonal antibodies (Gosztonyi and Ludwig 1984a; Ludwig et al. 1988, 1993). The sensitive unlabeled antibody-enzyme techniques demonstrated viral antigen in the nucleus, perikaryon and processes of nerve cells. Some neurons had only nuclear staining, while others had only cytoplasmic immunostaining. The appearance of the intranuclear antigen was variable. Frequently only one to six, relatively small (1–8 μm), round or ovoid aggregates could be seen (Fig. 2B); some of them were surrounded by a thin, light halo. The similarity of the aggregates to the Joest-Degen inclusion bodies was striking. These large aggregates occurred most frequently in the pyramidal cells of the hippocampus and in the large brain stem neurons. The nucleolus remained distinctly negative (Fig. 2B). It seems probable that virus-specific antigen forms aggregates, the size of which varies from the largest inclusion bodies down to submicroscopic, probably macromolecular, dimensions. Intranuclear inclusion bodies are sensitive to proteolytic enzymes: a 30 min treatment of tissue sections with trypsin eliminates the inclusions.

Cytoplasmic immunostaining is almost always diffuse and its intensity varies. Once the perikaryon contains antigen, it can soon be detected in all the processes, as disclosed by the positivity of the dendrites and the axon. With sensitive techniques, virus antigen can be shown to fill even the dendritic spines and the axons in CNS tracts, the terminal axonal networks, preterminal axonal varicosities and greater synaptic boutons. In the hippocampus, a band of diffuse immunostaining can sometimes be observed at low power examination both above and below the pyramidal cells, corresponding to the arborization area of the apical and basal dendrites of these neurons (Gosztonyi and Ludwig 1984a).

Besides neurons, astrocytes (Fig. 2C) and oligodendrocytes (Fig. 2D) may also contain virus-specific antigens (see Fig. 29 in Ludwig et al. 1985). These immunoreactive cells are restricted to areas where antigen-containing neurons and inflammatory infiltrations are present. Nuclei, perikarya and processes of glial cells may have antigens, but the staining of nuclei is inconsistent and almost always weaker than that of the cytoplasm. In the nuclei of astrocytes, virus antigen appears primarily in a fine granular or diffuse form, but occasionally, small, round aggregates with a light halo, resembling the Joest-Degen inclusion bodies, can also be found. The processes of affected glial cells are uniformly stained and their histological appearance is identical with the picture gained by specific glial impregnation techniques. The slender processes of oligodendroglial cells encircling and stretching along myelinated fibers are particularly impressive (Fig. 2D).

In situ hybridization shows BDV nucleic acids in great abundance in neurons, astroglial and oligodendroglial cells. In preparations with radioactive isotope-labeled probes, silver grains can be found above the nucleus, cell body and processes of these cells (Fig. 3) (Gosztonyi et al. 1991b, 1995). While antisense BDV probes label both the cytoplasm and nucleus, sense BDV probes, which detect viral genomic RNA, exhibit a clear emphasis above the nucleus (Fig. 3), in a similar way as in rats (Carbone et al. 1991a). In larger neurons, a strong focal concentration of hybridization signals can be seen in the vicinity of the nucleolus (Fig. 3, inset), corresponding to the site of the Joest-Degen inclusion bodies. The results seen in horse brains are similar to those observed in sheep, rabbits, rats, mice and monkeys (Gosztonyi et al. 1991b, 1995). The combination of in situ hybridization with immunohistochemical demonstration of viral antigens has

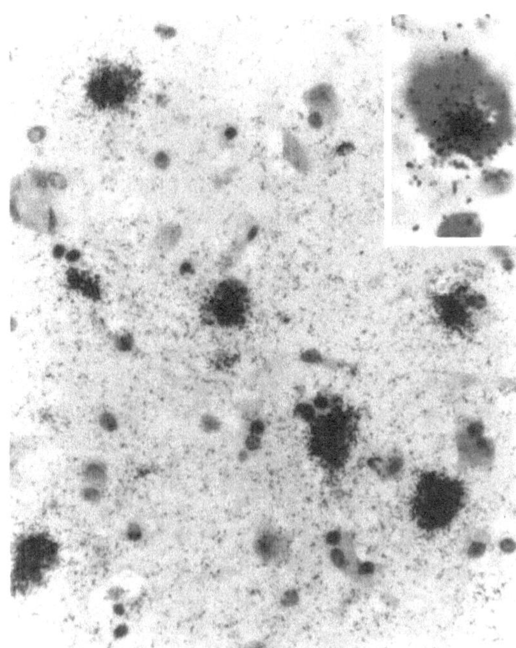

Fig. 3. In situ hybridization, cerebral cortex of the horse. Several neurons are heavily labeled. *Inset:* Hippocampal pyramidal cell labeled with a sense probe. Signals accumulate above a small nuclear area, most probably corresponding to a Joest-Degen inclusion body. Hematoxylin counterstain, x 477; *inset*, x 1130

shown that the overwhelming majority of CNS cells with hybridization signal also express viral proteins (Gosztonyi et al. 1995).

2.4 Electron Microscopy

With electron microscopy, no virus particles have yet been found in the brain of horses with BD. In the nuclei of neurons, nuclear bodies have been described in the involved regions (Bestetti 1976). These contain 55–90 nm electron-dense granules and spiraloid filaments in the center and a light halo and a microfibrillar capsule at the periphery. Based on comparative light and electron microscopic examinations these nuclear bodies were assumed to represent the Joest-Degen inclusion bodies (Bestetti 1976). In the cytoplasm, apart from occasional aggregates of fine filaments, no significant fine structural changes could be found (Gosztonyi and Ludwig 1984a; Ludwig et al. 1985).

2.5 Regional Distribution of Histological Lesions and Viral Products

Similar to other nonpurulent polioencephalitides, the lesions in BD are discontinuous and unevenly distributed along the neuraxis. Joest and Semmler (1911) defined first the preferential localization of the encephalitic process. They found the most severe inflammatory reaction in the olfactory bulb and gyrus, then in the caudate nucleus and hippocampus. The severity declined gradually in the caudal direction; only moderate changes were present in the spinal cord. JOEST and DEGEN (1911) pointed to the almost complete sparing of the cerebellum. SEIFRIED and SPATZ (1930) emphasized the strong involvement of the mesencephalon, the central gray matter with the oculomotor nuclei, the substantia nigra and the lamina quadrigemina. In the diencephalon, parts of the hypothalamus (infundibulum, tuber cinereum) and paraventricular parts of the thalamus are strongly affected. There is a severe inflammatory reaction in the caudate nucleus, while the putamen is frequently spared. Within the cerebral cortex, the hippocampus, olfactory gyrus and olfactory bulb are sites of severe inflammation. As a rule, the medio- and laterobasal cortical areas and the cingular gyrus are strongly affected whereas the upper lateral segments of the convexity are less severely involved (Fig. 1D). In the brain stem, the severity of inflammation declines caudal to the mesencephalon. The involvement of the vestibular and central cerebellar nuclei contrasts sharply with the sparing of the cerebellar cortex. In the spinal cord the severity of inflammation lags behind that of the cerebrum. The involvement of cranial and spinal sensory ganglia is very characteristic. The roots of the peripheral nerves are sometimes also involved.

Although BD is classified as a polioencephalitis, in the cerebral hemispheres, underneath the severely involved cortical areas, rather strong inflammatory infiltrates are seen, which occasionally surpass in severity those of the cortex (Fig. 1D). Therefore, it would be more correct to regard Borna encephalitis as a panencephalitis rather than as a polioencephalitis.

A comparative topographical study of the distribution of the inflammatory reaction and of virus-specific antigens disclosed that there is a close correlation between these two basic elements of the encephalitic process (GOSZTONYI and LUDWIG 1984a). Consequently, it is likely that the expression of viral antigens is the trigger of the cellular immune reaction. In the white matter, inflammatory infiltrations may be due to the presence of viral antigen in the axonal processes, astrocytes and oligodendrocytes.

2.6 Natural Infection of Sheep, Cats and Ostriches

In the sheep brain, BDV infection leads to a nonpurulent encephalitis which is similar to that of horses (BECK 1925; BECK and FROHBÖSE 1926). Adventitial inflammatory infiltrates can be found in both the gray and the white matter. Virus-specific antigens and nucleic acids can be demonstrated in the neurons. Inflammatory infiltrates are similar to those in horses, as shown in Fig. 1A.

A nonpurulent encephalomyelitis has been described in cats by LUNDGREN (1992). These cats had antibodies to the virus as well as BDV antigens and nucleic acid in brain (LUNDGREN and LUDWIG 1993; LUNDGREN et al. 1993; ZIMMERMANN et al. 1993).

Recently a paretic disease with high mortality has been described in ostriches held in farms in Israel (WEISMAN et al. 1993a,b; MALKINSON et al. 1993; see also

Chapter by Malkinson et al.). These animals had serum antibodies to BDV and BDV antigen in their brains (Malkinson et al. 1993). Morphological studies of these bird brains have not yet been completed.

3 Experimental Infections

Borna disease can be transmitted experimentally to a wide variety of laboratory animals. The clinical course, the virological and immunological parameters and the histopathology have been studied in great detail in chicken (Ludwig et al. 1973, Gosztonyi et al. 1983), mice (Kao et al. 1984; Rubin et al. 1993), rats (Nitzschke 1963; Hirano et al. 1983; Narayan et al. 1983a, b; Kao 1985; Carbone et al. 1987, 1989, 1991a, b; Ludwig et al. 1985, 1988), rabbits (Zwick et al. 1926, 1929; Pette and Környey 1935; Shadduck et al. 1970; Krey et al. 1979; Roggendorf et al. 1983), hamsters (Anzil et al. 1973; Blinzinger and Anzil 1973), tree shrews (Sprankel et al. 1978) and rhesus monkeys (Zwick et al. 1929; Pette and Környey 1935; Cervós-Navarro et al. 1981; Krey et al. 1982; Neubert 1984). In the majority of these animals, the infection results in an acute, subacute or more protracted disease, characterized by behavioral disorders, various neurological symptoms, progressive emaciation and cachexia leading to death. In most of these species, the nature of the histopathological alterations is similar to that of naturally infected horses and sheep, but there is a difference in the extension of the process. In natural disease, the histopathological alterations are bound to specific areas: olfactory centers, limbic system, retina, brain stem, while in experimental infections, the distribution of the changes is more diffuse.

In experimentally infected rats, the clinical course and histopathology vary with the age of the host at the time of inoculation, the genetic background of the host and the virus strain used for infection. Since the rat system has been so informative with respect to pathogenesis of the disease, rat infections are discussed here in greater detail.

3.1 Experimental Borna Disease of the Rat

3.1.1 Types of Experimental Disease

Rats infected as newborns with a low passage of the rabbit-adapted virus have high titers of BDV in their brains and high titers of anti-BDV antibodies in serum (Hirano et al. 1983; Narayan et al. 1983a, b; Kao 1985), but only subtle signs of altered behavior (Narayan et al. 1983a, b; Dittrich et al. 1989). After 12–16 months, they present progressive neurological deficits with paraparesis of the hind limbs and autonomic dysfunction (Gosztonyi et al. 1988; Ludwig et al. 1988).

If rats are infected as newborns with a high passage of the rabbit-adapted virus, they develop neurological signs in 1–4 weeks, similar to those seen in acute experimental Borna encephalitis. The disease is progressive and the majority of rats die within 1–4 months (KAO 1985; LUDWIG et al. 1985; KAO et al. 1995).

If weanling or adult rats are infected with a low passage of the rabbit-adapted virus, they develop the classic clinical picture of the acute experimental Borna encephalitis and die within 1–4 months. However, some rats infected in a similar way survive the acute disease and develop a marked obesity and exhibit aggressive behavior (KAO 1985; LUDWIG et al. 1985; KAO et al. 1983, 1990; GOSZTONYI et al. 1991a).

3.1.2 Persistent Infection

A persistent infection can be produced by intracerebral (i.c.), intraperitoneal (i.p.), intraneural, subcutaneous (s.c.) and intranasal (i.n.) inoculations. The latter probably simulates the natural mode of infection; furthermore, i.n. inoculation is followed by a progressive rostrocaudal spread of infection, so that in the same brain, early, intermediate and late phases of infection can be studied at the same time.

The i.n. infection is followed by a series of events, characterized by the progressive appearance of viral antigens and degenerative changes. Intriguingly, this occurs in the absence of cellular inflammatory infiltrations.

3.1.2.1 Virus Spread and Distribution of Virus-Specific Antigens. After nasal instillation of BDV, viral antigen appears first randomly in the olfactory neuroepithelium (Fig. 4A). Its presence, however, can be assessed only in the rostralmost part of the neuroepithelium. Neuroepithelial cells at the level of

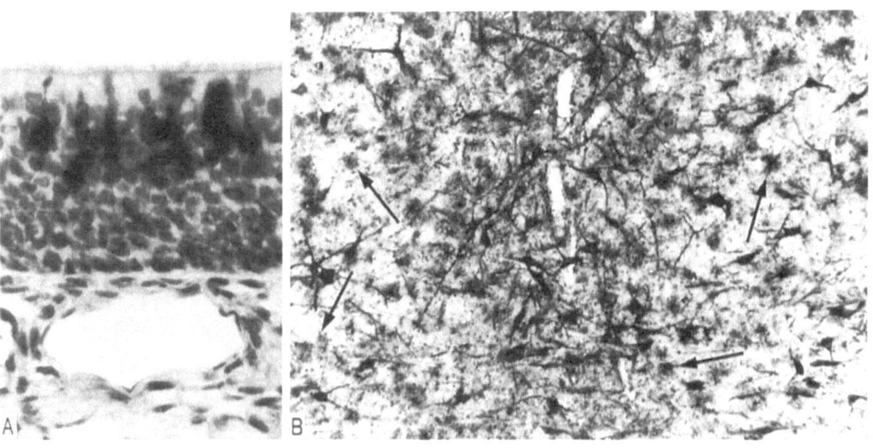

Fig. 4A, B. **A** Olfactory neuroepithelium of the rat. Several neuroepithelial cells contain BDV-specific antigen. x 565. **B** Paramedian region of the pontine tegmentum. The multipolar brain stem neurons are strongly stained with anti-BDV antibody in their perikarya and processes. Slight, patchy immunostaining of protoplasmic astrocytes (*slender arrows*). x 94

the olfactory bulb are completely negative at this stage. After about 4–6 days, strong expression of viral proteins is seen in the olfactory bulb. The positivity of the mitral cells is particularly prominent, but the granule cells of the bulb are also markedly affected. Viral antigens appear gradually and symmetrically in the secondary and tertiary olfactory centers: in the lateral olfactory nucleus, pyriform cortex, septum pellucidum. The early infection of neurons is the most spectacular phase of the spread of BDV: viral proteins, produced in the perikaryon, are transported to the neuronal processes. Immunohistochemistry at this stage reveals neuronal arborizations resembling neurons following Golgi impregnation (Fig. 4B).

Viral antigens are found in both the nucleus and the cytoplasm of infected cells. If examined with immunoelectron microscopy, infected cells have BDV-specific proteins in the paranucleolar chromatin. The nucleolus itself is negative at both light and electron microscopic levels. The Joest-Degen inclusion body probably represents viral protein-positive paranucleolar chromatin. By conventional staining it is acidophilic (Fig. 5A, B), by electron microscopy it is a round, granular or granulofilamentous formation, occasionally with areas of different densities (Fig. 5C).

Almost parallel with the appearance of viral proteins within the nucleus, there is a progressive shift of viral proteins produced in the perikaryon towards more peripheral segments of neuronal processes. When they reach the

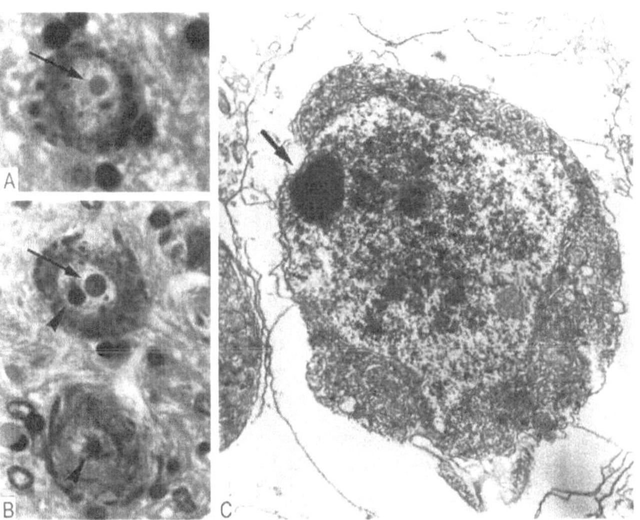

Fig. 5A–C. A Lumbar cord of the rat with experimental BD, ventral horn motoneurons. A Motoneuron with Joest-Degen inclusion body (*slender arrow*). **B** *Above;* Motoneuron with Joest-Degen inclusion body (*slender arrow*) and nucleolus (*arrowhead*). *Below:* Motoneuron with nucleolus only (*arrowhead*). x 970. **C** Olfactory cortex of the rat with BD. EM immunocytochemistry using a monoclonal anti-BDV antibody. Within the nucleus of a degenerating cortical neuron patchy labeling of smaller and larger granular aggregates. The largest aggregate (*arrow*) might correspond to a Joest-Degen inclusion body. x 16 000

preterminal and terminal axonal segments a diffuse staining of the neuropil is seen at immunohistochemical examination, with some similarity to nonspecific immunostaining.

Not all kinds of viral proteins are distributed evenly within a neuron. According to studies with defined antibodies (LUDWIG et al. 1993) the soluble (s-) antigen of BDV is more likely to be transported to peripheral segments of neurons, although it also appears in the nucleus. Some BDV antigens, detected by antibodies from chronically infected rabbits, appear to concentrate in the nucleus.

The pattern of virus spread following i.n. inoculation is consistent with the natural anatomical connections of the neurons first infected, i.e., virus spread is axonal and transsynaptic (transneuronal). From the secondary olfactory centers, the infection spreads to other parts of the limbic system, including the hippocampal formation. In the latter, the transsynaptic spread of the virus is particularly conspicuous. The hippocampal formation can be described as a chain of four neurons aligned linearly: (1) the perforant path as the main afferent system, (2) the neurons of the dentate gyrus establishing synaptic connections with (3) the pyramidal neurons of the CA4 and CA3 regions of the hippocampus, and, finally, (4) the pyramidal neurons of the CA1 region. The pyramidal neurons of the CA4 and CA3 regions are connected with the CA1 region through the Schaffer collaterals. All pyramidal neurons project through the fimbria hippocampi to the contralateral hippocampus (commissural pathways) and to other parts of the brain. Although direct connections from the entorhinal cortex to the CA3 region may result in early infection of a few pyramidal cells, the bulk of infectious virus arrives along the perforant path. Subsequently, the neurons of the dentate gyrus with their rich dendritic arborizations express BDV antigens and the infection proceeds along the mossy fiber system to the CA4 and CA3 pyramidal cells. The cell bodies and dendrites of the latter, together with their axons collected in the fimbria hippocampi, become strongly positive of BDV antigens (Fig. 6A). In the next phase of virus spread, BDV antigens appear in the CA1 region, in the neuropil between the basal and apical dendritic arborization of CA1 pyramidal cells, the stratum oriens and stratum radiatum. The most distal layer of the apical dendritic arborization of CA1 neurons, the stratum lacunosum-moleculare, remains distinctly free from viral antigen. This way, the antigen distribution shows a regular stratified pattern that consists of a narrow band corresponding to the basal dendrites (stratum oriens) and a broader band corresponding to the proximal part of the apical dendritic segments (stratum radiatum) (Fig. 6B). The cell bodies of the CA1 pyramidal cells and their dendrites remain distinctly free from viral antigen, and the apical dendrites appear as light vertical stripes within a diffusely immunostained neuropil (Fig. 6C). Evidently, preterminal axonal segments and axon terminals establishing synaptic contacts with the dendrites of CA1 pyramidal cells are stuffed with BDV antigen; constituents of infectious virus cannot traverse these synapses, in contrast to many other synapses they have passed on their way up to this point. This stratified distribution of viral antigen, bound strictly to a well defined anatomical structure, is an intriguing feature of BDV (GOSZTONYI and LUDWIG 1984b; LUDWIG et al. 1988; MORALES et al. 1988; GOSZTONYI

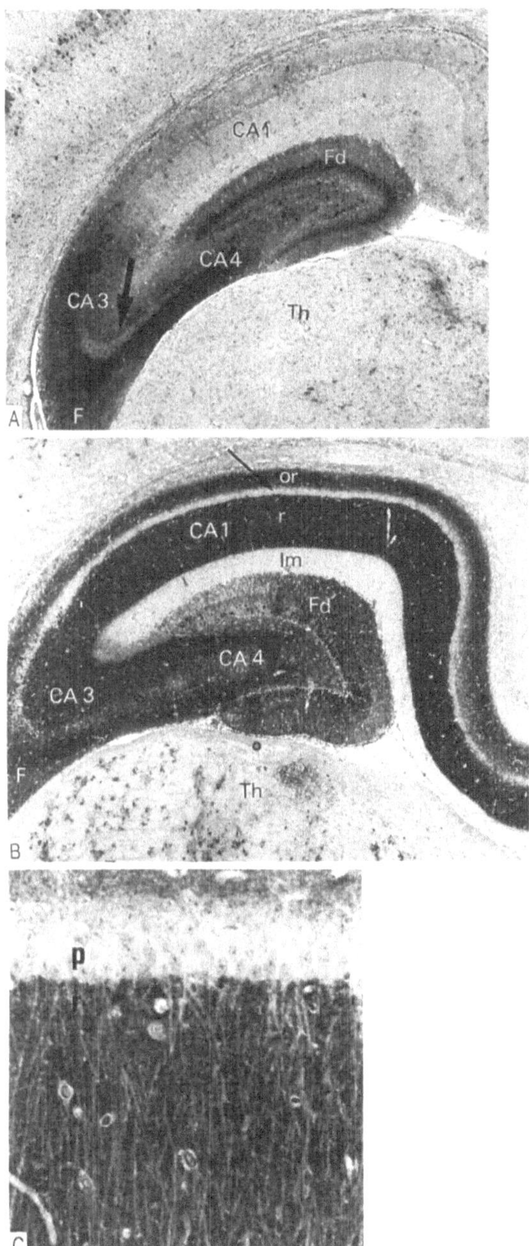

Fig. 6 A–C. A Spread of BDV in the rat hippocampus along a neuronal chain. Antigen is detected first in the dentate gyrus (fascia dentata, *Fd*), then the infection spreads along the mossy fibers (axons of dentate gyrus granular neurons, *arrow*) to the pyramidal cells of the *CA4* and *CA3* regions of the hippocampus. From the latter regions, infection spreads through the Schaffer collaterals of the pyramidal cells to the *CA1* region. In the latter, only a slight, initial, stratified positivity can be observed. *Th*, thalamus; *F*, fimbria hippocampi. Monoclonal anti-BDV antibody, APAAP technique. x 33. **B** Hippocampus of the rat in a later phase than shown in **A**. The *CA1* region shows a strong, stratified labeling of the stratum oriens (*or*) and stratum radiatum (*r*), while the stratum pyramidale (*slender arrow*) and the stratum lacunosum-moleculare (*lm*) are unlabeled. A few labeled neurons and astrocytes are detected in the thalamus (*Th*), *Fd*, fascia dentata, *F*, fimbria hippocampi. Immunostaining as in Fig. 6A, x 30. **C** *CA1* region of the hippocampus from the same animal as shown in **B**. The cell bodies and dendrites of pyramidal cells contain no BDV antigen. Antigen is found between the apical dendrites of the pyramidal cells of the neuropil, within synaptic boutons, terminal and preterminal axonal segments. Note the sharp border between the synaptic area of stratum radiatum (*r*) and the pyramidal cell layer (*p*). x 267

et al. 1993). After a longer period, viral infections break through this boundary and single CA1 pyramidal cells become infected.

The hypothalamus becomes infected rather early, the thalamus somewhat later. The neocortex comes to be infected progressively, starting partly from the cingular gyrus and partly from the pyriform cortex. In the meantime, through descending connections of the olfactory cortex, the infection reaches the brain stem. Numerous large multipolar brain stem neurons express BDV antigens, and protoplasmic astrocytes also turn positive (Fig. 4B). Later, this cell type becomes infected in other areas of the cerebrum as well. At the same time, central cerebellar nuclei, together with the dentate nucleus, become infected through their thalamic connections in a retrograde way. Starting from here, infection reaches the Purkinje cells (Fig. 7A). Expression of viral antigens in Purkinje cells is restricted to a rather short period. In about 2–3 weeks they are completely cleared from viral proteins (Ludwig et al. 1988). Parallel with this change, astrocytes of the cerebellar cortex, the Bergmann glia cells, become infected, and their processes show a diffuse immunostaining in the molecular layer (Fig. 7B). Many months later, in some rats, the positivity of the Bergmann glia cells subsides. In the meantime, the small neurons in the granule cell layer become positive. Their axons, which project to the molecular layer of the cerebellar cortex, are filled with viral antigen and contrast sharply with the unstained dendrites of the Purkinje cells, with which they form synapses (Fig. 7C).

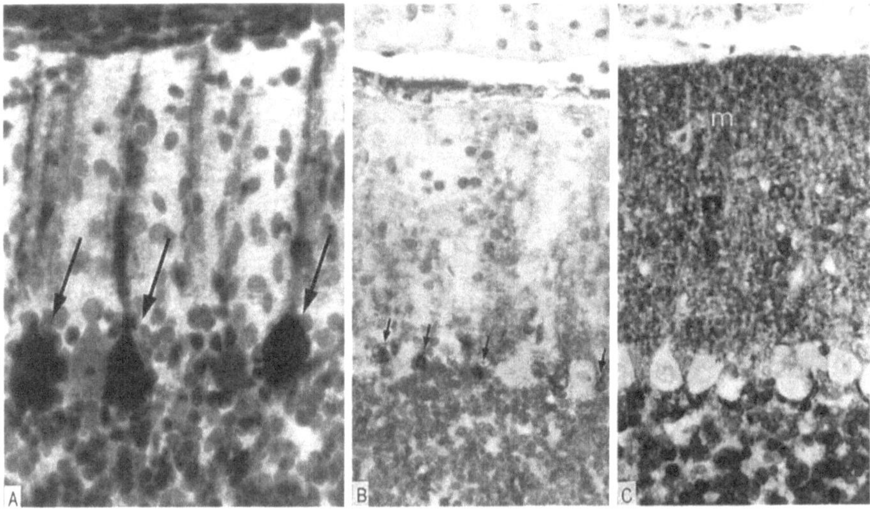

Fig. 7A–C. Three types of BDV antigen distribution in the cerebellar cortex in three phases of persistent infection. **A** Early phase. BDV antigen is present only in a few Purkinje cells (*slender arrows*). **B** Intermediate phase. The Purkinje cells are free from virus antigen that is found now in the cell bodies and processes of Bergmann glia cells (*small arrows*). **C** Late phase. Small granular neurons underneath the Purkinje cell layer are positive for BDV antigen and viral antigen in their processes leads to a diffuse immunostaining of the molecular layer (*m*). Purkinje cells and their dendrites are negative for BDV antigen. **A** x 377; **B** and **C** x 218

The phasic expression of BDV antigens is easily observed in the cerebellar cortex, due to its well defined structure. It is likely that expression of BDV antigen is phasic in other cerebral areas as well, but this cannot be verified by light microscopy.

There is, however, another approach which provides evidence for the phasic expression of BDV antigens (GOSZTONYI et al. 1993). At electron microscopy, BDV antigens can be seen in various compartments of an infected cell. If the immunostaining of the nuclei and perikarya are comparatively examined, three patterns are observed: Viral antigen is expressed (1) only in the cytoplasm; (2) in the cytoplasm and the nucleus; (3) only in the nucleus (Fig. 8A, B, and C). These patterns can be interpreted as a sequence of events: (1) viral proteins are synthesized in the cytoplasm, (2) viral proteins are transported to the nucleus; (3) viral protein synthesis is repressed, the protein already produced is transported into the nucleus, (4) BDV antigens are eliminated or their antigenic structure is altered.

Neural cells other than neurons and astrocytes are also infected. Infection of oligodendrocytes ensues rather late, several months after primary infection. Their slender processes become filled with viral antigen, so that their immunostaining resembles specific histological impregnations of this cell type. Within 6–12 months following primary infection, ependymal cells, choroid plexus epithelial cells, and cells of the pineal gland can express BDV antigens. Constituents of the cerebral vessels, connective tissue and hematogenous cells never express BDV-specific proteins.

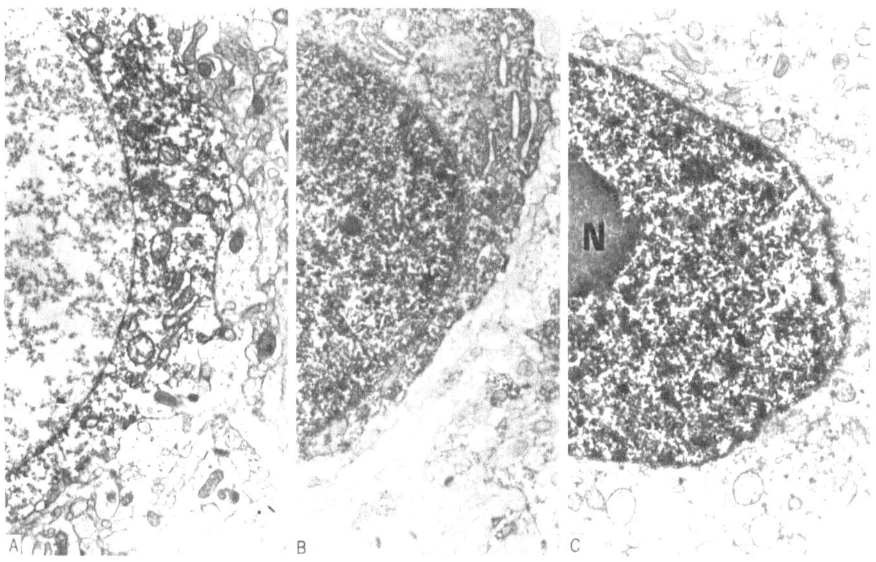

Fig. 8 A–C. Electron microscopic demonstration of BDV antigen in neurons. **A** Only the cytoplasm is labeled. **B** Both nucleus and cytoplasm are labeled. **C** Only the nucleus is labeled. The nucleolus (*N*) remains negative. **A**, **B** and **C**, x 8000

Some 2–6 weeks after viral antigen first appears in the brain, virus spreads centrifugally to the spinal cord, cranial and spinal radices and nerves. Viral antigen appears rather early in the trigeminal nerve and ganglion. The olfactory neuroepithelium becomes strongly immunoreactive, not only in its rostralmost part but diffusely in all its extensions. In the retina, virus antigen appears first in the ganglion cell layer, and spreads gradually along its neuronal chain, until it reaches the photoreceptor layer (LUDWIG et al. 1985; CZUB 1988). In the spinal cord, infection spreads earlier along the ascending sensory systems, probably by the retrograde axonal transport. In the lumbar cord, in the early phase of spinal cord involvement, the posterior horns, sensory roots and sensory ganglia are positive

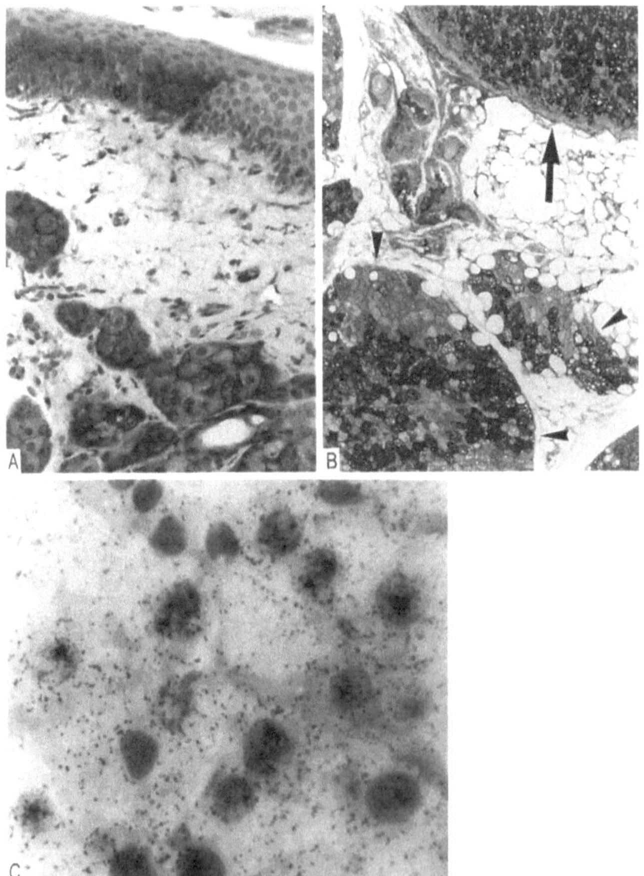

Fig. 9 A–C. **A** Skin of the upper lid of a persistently infected rat. The sebaceous glands and the squamous epithelial cells express BDV antigen. x 232. **B** Adrenal gland (*arrow*) and brown fat (*arrowheads*) of a rat persistently-infected with BDV. In both tissues only selected cells express BDV antigen. x 91. **C** Lacrimal gland of a rat persistently infected with BDV. In situ hybridization with a sense probe. Signals are concentrated above the nuclei of the secretory epithelial cells. x 1131

for viral antigen, while the descending systems, the anterior horns and ventral roots are still negative.

Late in infection, BDV antigen appears in the axoplasm of the peripheral nerve fibers of all tissues and organs. During this centrifugal spread, Schwann cells and satellite cells of sensory and autonomic ganglia (amphicytes) also express viral antigen, but this expression is far less constant than in the oligodendrocytes, the sheath cells of central axons (CARBONE et al. 1989; GOSZTONYI et al. 1993).

As a consequence of the centrifugal spread of viral constituents, BDV is presented to many tissues and visceral organs (GOSZTONYI et al. 1987, 1993). In spite of this fact, several months pass until BDV breaks through the barrier to nonneural tissues. Late in infection, BDV antigens appear in the lacrimal, salivary and sebaceous glands (Fig. 9A), in the squamous epithelium of the skin (Fig. 9A), in the thyroid, in the anterior (but also intermediate and posterior) pituitary, in the exocrine and endocrine part of the pancreas, intestinal epithelia, adrenal cortex (Fig. 9B) and medulla, in a few cells of the liver and kidney and in the brown fat tissue (Fig. 9B). In general, it can be ascertained that (with the exception of brown fat and perhaps some smooth and striated muscle cells) all ectodermal tissues/epithelial cell types are permissive to BDV. Until recently, it was not possible to determine whether these nonneural cells in fact replicate BDV, or only take up BDV antigens presented to them by the axonal flow. The demonstration of BDV nucleic acids (both sense and antisense RNAs) by in situ hybridization in the lacrimal gland of the rat has demonstrated that active BDV replication takes place in nonneural cells (GOSZTONYI et al. 1995) (Fig. 9C). These findings in experimental rat infections correlate well with nucleic acid and antigen findings in extraneural organs of naturally infected horses (ZIMMERMANN et al. 1993; DÜRRWALD 1993; Bode, unpublished results).

3.1.2.2 Virus-Induced Parenchymal Damage. Since in the early intermediate phase of persistent infection no obvious neurological deficits can be documented in the rat, it was believed that this condition is a tolerant infection, in which the presence of BDV in neurons does not really interfere with their function or structural integrity. However, elaborate clinical study reveals neurological and neuropsychological deficits, which become progressively prominent in the later phases of infection (NARAYAN et al. 1988a, b; DITTRICH et al. 1989).

Histologically, the majority of neurons that harbor viral antigens have a normal structure at light microscopic examination. However, there are CNS areas that undergo severe degeneration in the course of persistent infection. The rostral half of the dentate gyrus is most severely and most regularly affected (LUDWIG et al. 1988; CARBONE et al. 1991b) (Fig. 10A). Neuronal degeneration begins in the inner layers of the structure, and those layers become progressively thinner. Degenerating dentate gyrus neurons express BDV antigens. In the late phase, only a thin row of cells can be found, or the structure disappears completely. Sometimes the ventral blade of the structure is more severely affected than the dorsal one. A marked reactive astrocytosis develops in parallel with the disappearance of dentate gyrus neurons (Fig. 10B).

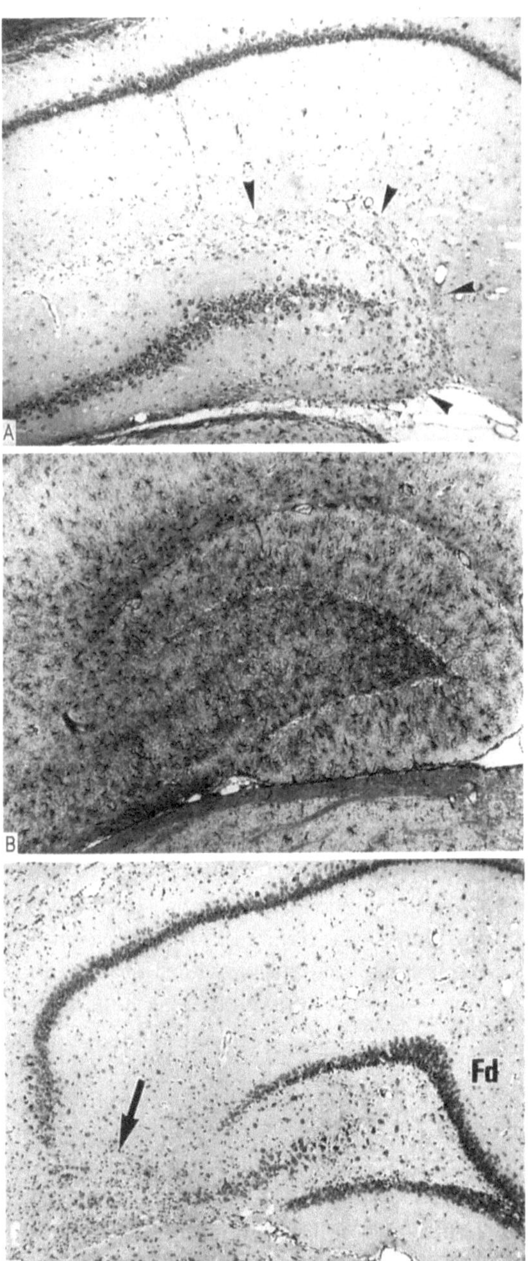

Fig. 10 A–C. A Hippocampus of a rat persistently infected with BDV. The neurons of the dentate gyrus have completely been destroyed, and replaced by reactive astrocytes (*arrowheads*). Nissl stain, x 55. **B** Severe reactive astrocytosis after virus-induced destruction of the dentate gyrus. GFAP immunostaining, x 57. **C** Virus-induced partial destruction of the CA3 region of the hippocampus (*arrow*) in the course of persistent infection with BDV. The dentate gyrus (Fd), and the CA4 and CA1 regions are preserved. Nissl stain, x 68

Less frequently, the CA3 and CA4 regions undergo degeneration. The pyramidal cells become pyknotic, disappear, and are replaced by macrophages and a reactive astrocytosis (Fig. 10C).

The retina is also vulnerable to BDV infection and virtually all persistently infected rats have retinal damage in the later phase of the disease (HIRANO et al. 1983; LUDWIG et al. 1988). First, all the layers of the retina are inundated by the virus and express viral antigen (Fig. 11A). Later, beginning with the inner granular layer, the neurons of the retina degenerate and the structure becomes thinner (Fig. 11B, C). The animals are blind at this phase and many of them develop cataracts (KAO 1985; LUDWIG et al. 1985). Eventually the optic nerve atrophies and becomes gliotic.

After about 8–10 months, neuronal degeneration extends to other cerebral areas as well. This process does not destroy entire structures and is rather random. Late in the course BDV antigen-negative neurons can be found with almost complete lysis of the cytoplasm and relative sparing of the nucleus. The result of this diffuse neuronal degeneration is a decrease in brain mass leading to internal hydrocephalus (LUDWIG et al. 1985).

In spite of widespread infection of astrocytes, no obvious cytopathic changes can be found in this cell type, and their infection does not compromise their reactive capacity.

Oligodendrocytes are profoundly damaged by BDV infection, with vacuolation of tracts, mainly in the brain stem but also in the intracerebral portions of the radices of cranial nerves (Fig. 12A). In the spinal cord, the BDV-induced oligodendrocyte damage leads to a severe vacuolation of the ventrolateral and posterior funiculi and to vacuolar myelopathy (GOSZTONYI et al. 1988). Perhaps the most severe damage is found in the cerebellar white matter where a circumscribed

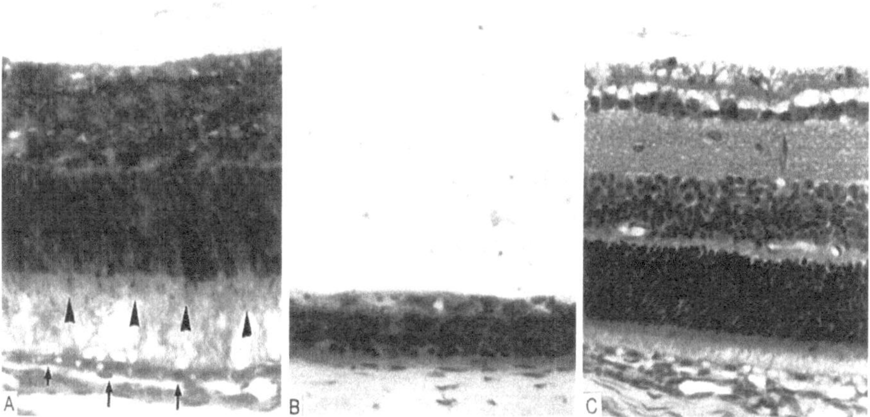

Fig. 11 A–C. A Retina of a rat infected with BDV. Antigen in the photoreceptor layer is only random, and is infrequent (*arrowheads*); other layers have diffuse expression of viral antigen. The majority of pigment epithelial cells (unpigmented in the albino rat) also expresses BDV antigen (*small arrows*). **B** Severely destroyed retina of a rat persistently infected with BDV. Hematoxylin stain. **C** Normal retina of a rat. H & E stain. **A**, **B**, and **C**, x 218

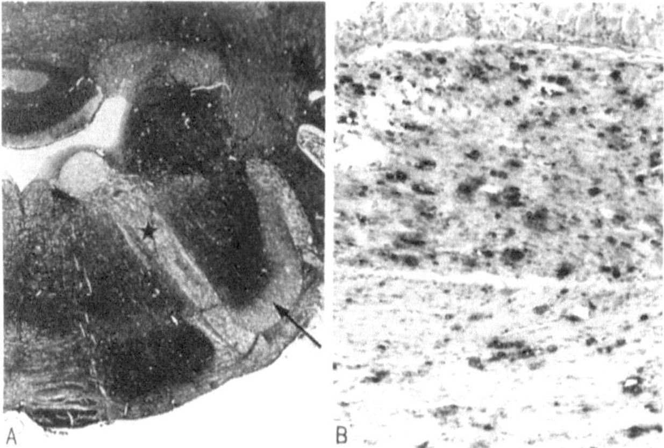

Fig. 12 A, B. A Lower segment of the pons. Some intracerebral tracts, such as the intrapontine radix of the facial nerve (*asterisk*) and the tractus descendens of the trigeminal nerve (*slender arrow*), are severely vacuolated. Klüver-Barrera stain, x 15. **B** Trigeminal nerve of the rat persistently infected with BDV. Strong infiltration of the endoneurial space with macrophages. Immunostaining with an anti-macrophage antibody (ED 1), x 88

spongy dissolution of the white matter is associated with a reactive fibrillary gliosis.

3.1.2.3 Reactive and Inflammatory Changes. The reactive astrocytosis accompanying the focally emphasized degenerative changes has already been described. The random degeneration of neurons and the vacuolar change of the tracts is also accompanied by reactive astrocytosis and fibrillary gliosis.

An inflammatory reaction does not precede the degenerative changes. A certain degree of macrophage activation can be seen in the early phases of persistent infection. The macrophages are found mainly in areas of high BDV antigen concentration; the latter seem to attract monocytes/macrophages. This is particularly obvious in CNS tracts, where viral antigen is axonally localized. In the intermediate phase only a few macrophages are found, but in the late phase, when parenchymal damage is in the foreground, macrophages appear again and seem to remove breakdown products. This process is particularly obvious in the trigeminal nerve (Fig. 12B).

There are a few exceptions to this general rule. An intense inflammatory reaction consisting of lymphocytes and monocytes may be seen in the pituitary stalk and the infundibulum. Similar inflammatory infiltrates can be found in the pineal gland, if there is an expression of BDV antigens. Both of these areas are external to the blood-brain barrier. Furthermore, in extraneural tissues infected with BDV such as the lacrimal and salivary glands, several compact lympho-monocytic infiltrates can also be found.

3.1.2.4 Electron Microscopy. Electron microscopic studies have been done in various phases of persistent infection and in various regions of the rat brain. Virus particles have never been found, even when the frontline of the progression of

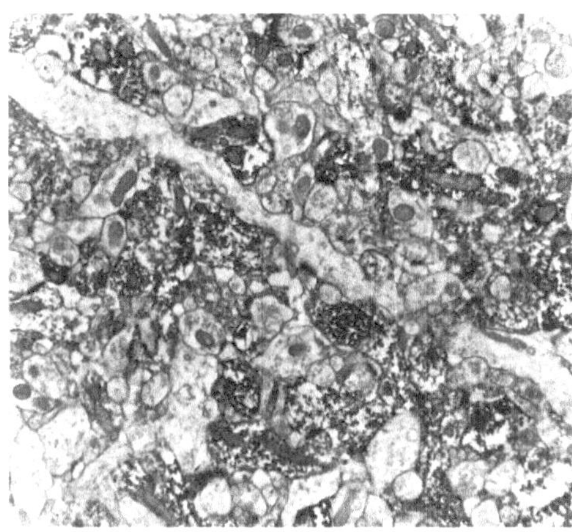

Fig. 13. Neuropil of a rat with persistent BDV infection. Electron microscopic demonstration of the s-antigen of BDV. Approximately half of the neuronal processes are diffusely positive for the s-antigen. x 8000

infection has been examined. Infected neurons present a more or less pronounced loss of ribosomes and rough surfaced endoplasmic reticulum. The nuclear changes have been described above. In the late phase, some intranuclear crystalloids and cytoplasmic stacks of filaments can be found; we regard these as non-specific changes, since they have also been described in old, otherwise healthy rats (Brion et al. 1982).

Immunoelectron microscopy has not facilitated identification of virus particles or virus-specific structures. The cytoplasm and processes of infected cells are filled with viral antigens. In the nucleus, the paranucleolar chromatin becomes positive first, followed later by the remainder of the nuclear chromatin (Fig. 8B, C). In the initial phase of neuronal infection, the immunostaining of the cytoplasm is focal. In the neuropil, the axonal processes and terminals of infected neurons also stain diffusely. BDV antigen-positive and -negative processes lie side by side (Fig. 13). Infected astrocytes also exhibit diffuse immunostaining. Positive immunostaining has also been demonstrated electron microscopically in the receptor cells of the olfactory neuroepithelium and in Schwann cells (Gosztonyi et al. 1993).

3.1.3 Hyperacute Infection

If newborn rats are infected with high passage, rat-adapted virus, they develop a severe, progressive neurological disease and die within 1–6 weeks (Kao 1985; Ludwig et al. 1985; Kao et al. 1995). Rats that die early have devastating neuronal degeneration. Nuclear chromatin is marginated (Fig. 14), and the oval or round center is filled with an eosinophilic material, sometimes surrounded by a clear zone. Electron microscopically, the eosinophilic material is finely filamentous. Neurons and neuropil have diffusely distributed viral antigens; inflammation is completely absent. Only a few scattered macrophages can be demonstrated by immunohistochemical techniques.

In rats with a somewhat more protracted course, a series of events can be seen histopathologically and immunohistochemically that are basically identical to those in the persistent form of the disease. However, the speed of develop-

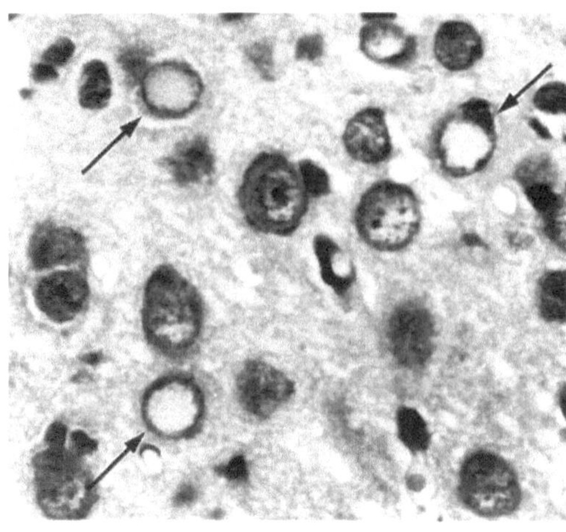

Fig. 14. Infection of the rat with high passage BDV. Severe lytic changes, not only in the cytoplasm of neurons, but also in some nuclei (*slender arrows*). H & E stain, x 696

ment of these changes is accelerated. No inflammatory reaction can be detected, except for macrophage activation of varying intensity.

3.1.4 Acute/Subacute Infection

Following i.n. infection of weanling or adult rats, the spread of infection and the distribution virus-specific antigens is similar to that seen in persistent infection. The process is, however, accompanied by an intense inflammatory infiltration, which follows roughly the distribution of CNS cells expressing BDV antigens. The inflammatory cells are located in the leptomeningeal and adventitial spaces, and diffusely in the neuropil. Neuronophagias are seen only infrequently. A recent analysis using leukocyte-specific monoclonal antibodies disclosed that the composition of the leptomeningeal and adventitial infiltrates is very similar; they consist of helper and cytotoxic/suppressor lymphocytes, monocytes and very few B cells. The diffuse tissue infiltrates are composed almost exclusively of monocytes/macrophages; T lymphocytes rarely enter the parenchyma (Petrov 1993).

3.1.5 Intracerebral, Intraocular and Peripheral Inoculations

Following i.c. inoculations expression of BDV antigens begins at multiple sites, presumably because virus spreads throughout the CSF. The ventral blade of the dentate gyrus and the CA3 region of the hippocampus are infected early, because they lie close to the leptomeningeal space of the hippocampal fissure and are particularly susceptible to BDV infection.

After intraocular inoculation, BDV spreads transsynaptically according to the rich anatomical connections of the optic system (Czub 1988).

After BDV inoculation into the lower limb or into the sciatic nerve the infection spread also transsynaptically in caudo-rostral direction within the spinal cord (Pette and Környey 1935; Carbone et al. 1987).

3.1.6 Obesity Syndrome

The neuropathological picture of this syndrome depends in part upon the phase of the disease when the brain is examined. Based on a study of 40 rat brains, the picture can be summarized as follows: There is only slight cellular inflammatory infiltration. In the earlier phases, BDV antigens are present in many neurons, but

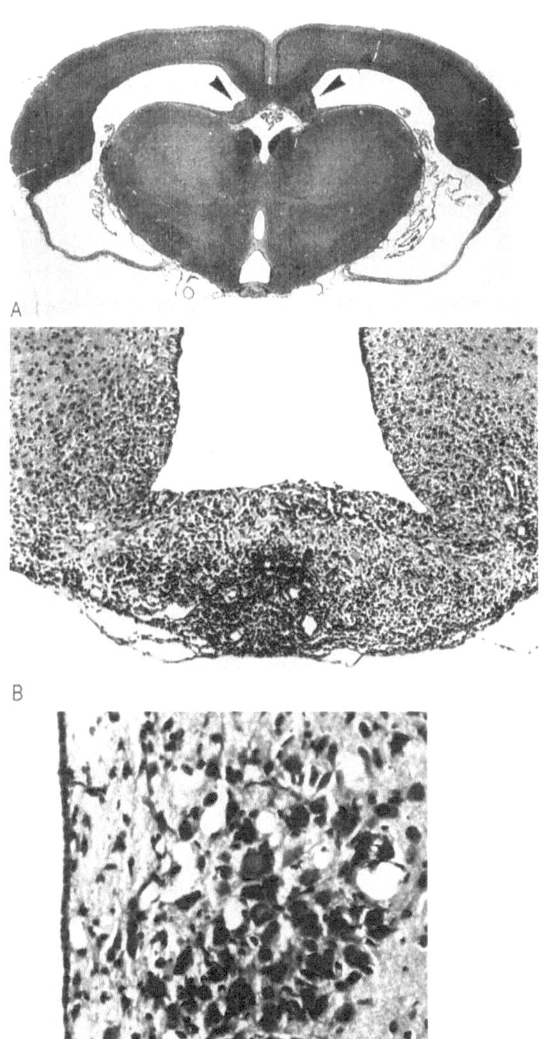

Fig. 15 A–C. BDV induced obesity syndrome. **A** Severe, almost complete, involution of the hippocampus, leaving only a rudiment behind (*arrowheads*). H & E stain, x 8. **B** Granuloma-like lymphomonocytic inflammatory infiltration in the infundibulum of the hypothalamus. H & E stain, x 96. **C** Vacuolar degeneration of neurons in the paraventricular nucleus of the hypothalamus. H & E stain, x 297

the number of antigen-positive cells decreases progressively; after 18–24 months hardly any BDV-positive neurons or glial cells can be found. Degeneration of neurons is also progressive, particularly in the dentate gyrus, the CA4 and CA3 regions and the retina. By 12–18 months there is a progressive involution of the whole hippocampal formation, so that only a small, shrunken rudiment remains under the corpus callosum, near the midline (Fig. 15A) and the fimbria hippocampi becomes extremely narrow. In a few animals, involution of the entorhinal cortex accompanies the dissolution of the hippocampus (Fig. 15A). A profound hydrocephalus ex vacuo accompanies this atrophic process. Due to loss of neurons in the neocortex, this structure becomes moderately narrow. In the cerebellum, foci of spongy degeneration develop. The neuronal degeneration is followed by marked reactive astrocytosis.

The hypothalamus has been examined on serial sections in the majority of our series of 40 obese rats. A lymphomonocytic inflammatory infiltration in and around the infundibulum was a prominent and constant finding (Fig. 15B). Occasionally, the inflammation had a granulomatous character. In a few cases, adventitial cellular infiltrates were found in the hypothalamus remote from the infundibular region. Several rats had a symmetrical, slight to moderate vacuolar degeneration of the neurons of the paraventricular nucleus (Fig. 15C). Visceral organs were remarkable for the enormous deposition of fat tissue and hyperplasia of the Langerhans islets. Obese rats showed marked increases in serum triglyceride levels and a moderate increase in blood glucose (Kao 1985; Kao et al. 1990).

3.2 Experimental Borna Disease of Mice

Following adaptation of BDV to the mouse, the neuropathology of this experimental disease has been studied in great detail (Kao et al. 1984; Rubin et al. 1993). The pathology is similar to that seen in persistently infected rats. Inflammation is minimal or absent.

3.3 Experimental Borna Disease of Rabbits, Hamsters, Tree Shrews, Monkeys and Chicken

The disease in these species is similar both clinically and histopathologically to natural host disease. The main difference is that in these experimental infections the inflammatory process is more diffuse.

BDV-specific antigens were found in rabbits, monkeys and chicken in neurons and glial cells. In situ hybridization studies were performed also in rabbits, rhesus monkeys and chicken (Gosztonyi et al. 1991b, 1995). Both sense and antisense viral RNAs could be demonstrated abundantly. Genomic RNA, as in the horse (see Fig. 3) and in the rat (Carbone et al. 1991a), was mainly present in the nuclei, with a paranucleolar emphasis, while mRNA could be demonstrated both in the nucleus and cytoplasm.

4 Pathogenesis

4.1 Virus Spread

In natural infections, the portal of entry of BDV is likely to be the olfactory neuroepithelium. Horses have a severe and early inflammatory reaction (JOEST and SEMMLER 1911). Experimental i.n. infection has been successful in rabbits, rats and mice (ZWICK and SEIFRIED 1927; ZWICK et al. 1926, 1929; MORALES et al. 1988; GOSZTONYI et al. 1985). In rats, viral antigen is found in the olfactory neuroepithelium as early as 4 days after i.n. infection. In the early phase of infection, spread of BDV within the nervous system is axonal and transsynaptic (transneuronal) (LUDWIG et al. 1985, 1988; CARBONE et al. 1987; GOSZTONYI et al. 1984, 1993). The neuronal system first infected at the portal of entry determines the spread of infection according to its natural connections: i.n. inoculation leads to infection along the olfactory and limbic systems; intraocular inoculation results in infection along the connections of the optic nerve; foot pad and limb intraneural inoculations result in spread of infection along the ascending systems of the spinal cord. This transsynaptic spread can be anterograde and retrograde: anterograde, if the cell body (perikaryon) of the neuron is infected first, and retrograde, if, e.g., a peripheral receptor area or a peripheral nerve is infected first. The two mechanisms are basically different, but BDV spreads within the CNS in both directions with great ease.

The mode of spread implies that viral genetic material must cross the synapse. This event is closely connected with the difficulty that virus particles have never been demonstrated in the brain in BDV infection. The conventional mode of virus spread requires full replication cycles, production of complete virus particles, their discharge into the extracellular space and entry into an uninfected cell. We tried repeatedly to find virus particles with the electron microscope in the frontline of infection, i.e. in areas where the infection in fact crosses a well defined synapse; but virus particles could never be visualized. Though, immunoelectron microscopy shows the appearance of viral antigens and their diffuse distribution in the infected cell. To resolve this problem, we proposed (GOSZTONYI et al. 1993, 1994) that the genetic material of BDV, its RNA alone or its ribonucleoprotein, crosses synapses in its macromolecular and morphologically unperceivable form. A similar mechanism is at work at the frontline of spread of another negative strand RNA virus, rabies virus (GOSZTONYI 1978, 1979, 1986, 1994; GOSZTONYI et al. 1990, 1993). Molecules and macromolecules are carried easily under physiological conditions axonally and transsynaptically (ORIOLI and STRICK 1989); viral macromolecules may use the same mechanism for their neuronal spread.

4.2 Borna Disease Virus and Neurotransmitter Receptors

Entry of a virus into a cell may be mediated by specific receptors. Though specific neurotransmitter systems are affected by infection with BDV (LIPKIN et al. 1988), there are presently no data to indicate that the virus uses neurotransmitter systems for infection. Intriguingly, some synaptic fields are spared, e.g., the inhibitory, GABAergic system of the hippocampus, terminating on the cell bodies of pyramidal cells (Gosztonyi and Ludwig 1984b; Gosztonyi 1985; Gosztonyi et al. 1994).

4.3 Borna Disease Virus Antigens in the Hippocampal Formation: Their Affinity for Aspartate and Glutamate Receptors

Borna disease virus antigens have a specific, stratified distribution in the hippocampus that can be correlated with aspartate and glutamate neurotransmitter systems: BDV antigens are found in the stratum oriens and stratum radiatum;

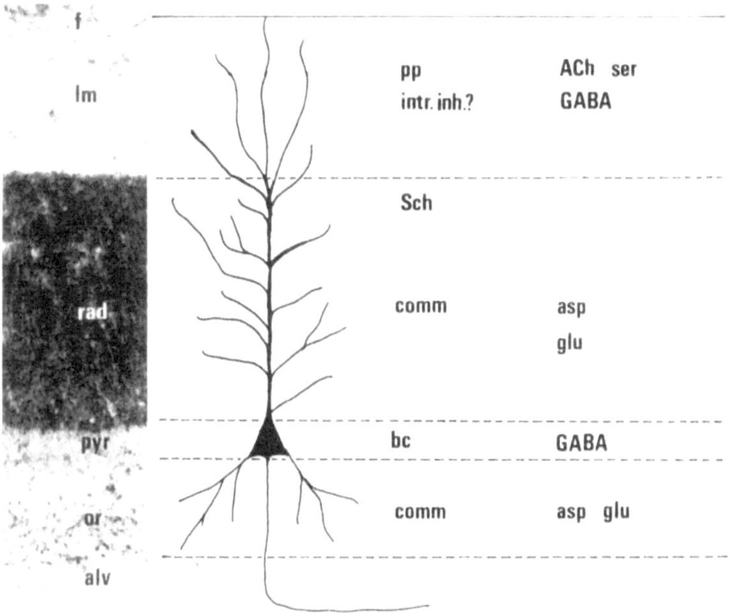

Fig. 16. Comparative demonstration of the distribution of BDV-specific antibody in the CA1 region of the hippocampus in a rat persistently infected with BDV (*left*), the pattern of dendritic arborization of hippocampal pyramidal neurons (*middle left*), the patterns of termination of afferent systems in various layers of the hippocampus (*middle right*) and the putative transmitters used by the terminals of the afferent systems (*right*). *f*, hippocampal fissure; *lm*, stratum lacunosum moleculare; *rad*, stratum radiatum; *pyr*, stratum pyramidale; *or*, stratum oriens; *alv*, alveus; *pp*, perforant path; *intr. inh*, intrinsic intrahippocampal connections; *Sch*, Schaffer collaterals; *comm*, commissural connections; *bc*, basket cells; *ACh*, acetylcholine; *ser*, serotonin; *GABA*, γ-amino butyric acid; *asp*, aspartate; *glu*: glutamate

however, no antigens can be demonstrated in the pyramidal cell layer, in the dendrites of the pyramidal cells or in the stratum lacunosum moleculare. The two BDV antigen-positive layers are sites for termination of the Schaffer collaterals of the ipsilateral CA4 and CA3 pyramidal cells and the axons of the contralateral CA4 and CA3 pyramidal neurons through association pathways along the fimbria hippocampi. The terminals in the two layers use aspartate and glutamate as transmitters (Fig. 16).

4.4 Cell and Tissue Tropism of Borna Disease Virus: Elective Vulnerability

Nonneuronal cells of the nervous system are also permissive for BDV (astrocytes, oligodendrocytes, ependymal and plexus epithelial cells, Schwann cells and perineurial satellites of sensory and autonomic ganglia). The mechanism for virus uptake in neuronal and nonneuronal cells is not known. Many glial cells express various neurotransmitter receptors; perhaps BDV constituents are bound to and taken up by them. In persistently infected animals, usually after 8–12 months, BDV appears in various extraneural tissue and organs as well.

These observations call into question the view that BDV is a strict neurotropic pathogen. Nevertheless, the virus replicates primarily in neurons and secondarily in glial cells. Extraneural organs and tissues appear to become infected only if virus is delivered via the peripheral axons for a long period of time. Even then, infection seems to be restricted to epithelial tissues. The one mesodermal tissue that is frequently infected is brown fat, a tissue that is permissive in many other viral infections.

Until the introduction of immunocytochemistry and in situ hybridization in neurovirology, the terms neurotropism, special neurotropism and elective vulnerability, or pathoclisis have been almost synonymous (Gosztonyi 1985). The methods mentioned above have shown that many neural cells may be infected, express viral proteins and still function normally. BDV has been regarded to be a noncytopathic virus, because in tissue culture it has no effect on cell growth or survival. In vivo, however, BDV can be cytopathic. The dentate gyrus, the CA3 and 4 regions of the hippocampus and the retina are especially vulnerable and undergo virus-induced degeneration in a relatively short time. Many other neurons shows signs of degeneration after a longer period of infection. BDV apparently interferes, at least partially, with the normal functioning of oligodendroglial cells as well. This manifests itself in the vacuolation of CNS tracts. White matter pathology, particularly vacuolar myelopathy, may contribute significantly to the development of neurological disease. Beyond doubt, there are important host factors determining the development of these elective lesions. What makes these regions and cell types vulnerable to BDV infection is not known.

Tropism and pathogenesis are complex in BD. The view that BDV has affinity for certain brain areas (hippocampus, pyriform cortex, entorhinal cortex, or the limbic system as a whole, and the brain stem) may need revision. The importance of the time factor has been stressed earlier (Környey 1939, 1954): in the early phases of a viral encephalitis the portal entry is decisive in determining the distribution of the process. The factors determining intrinsic vulnerability become apparent only later, when the agent has had the chance to reach all brain areas. In natural BD, the predilection to involve the olfactory, namely, the limbic, system may only reflect the initial spread of the agent following nasal infection. Since the majority of naturally infected animals die early through euthanasia or involvement of the brain stem, it may appear that BDV specifically targets these structures.

In discussing tropism it is important to distinguish between site-specific targeting by the virus, which may result in persistent, noncytopathic infection, and **elective vulnerability** or **pathoclisis**. The latter expression should be used primarily to characterize specific vulnerability of certain neuronal systems in neurodegenerative disorders (Vogt and Vogt 1920). In BD, degeneration of hippocampal structures and the retina may be regarded as an example of elective vulnerability.

The elective vulnerability has an important role in pathogenesis during persistent infection. Learning deficiency is a prominent finding in persistently infected rats (Dittrich et al. 1989). This has its explanation in the extensive and severe virus-induced degeneration of the dentate gyrus. The perforant path, terminating on dentate gyrus granule cells, is the principal afferent system to the hippocampus. Destruction of the dentate gyrus results in deafferentation of the hippocampus and compromise of two of its main functions, learning and memory. Whether this deafferentation contributes also to the involution of the hippocampus seen in the obesity syndrome, is unclear.

In the earlier phase of infection, it is possible that viral infection may interfere with synaptic function in the stratum oriens and stratum radiatum of the hippocampus, resulting in partial deafferentation of the CA1. Such a mechanism could account for the severe EEG disturbances observed in BDV-infected rabbits (Avdaloff et al. 1981; Gierend 1982).

4.5 Phasic Expression of Viral Proteins

An interesting feature of BDV infection of the nervous system is the phasic expression of viral proteins. It is unclear whether this control is determined by BDV or the host cell. Some neurons, e.g., Purkinje cells, seem to have only a single early period of virus protein production, while in other cells, repeated periods of antigen expression and repression may alternate. Periods of repression of viral protein production may favor host cell recovery.

4.6 Obesity Syndrome

Escaping lethal disease at the cost of developing an obesity syndrome is a special feature of experimental BD which emphasizes the importance of the elective vulnerability. Why a few rats survive the acute disease is unclear. For the explanation of the obesity the neuropathologic study can offer some possibilities. Studies of the hypothalami of 40 obese animals disclosed strong inflammatory infiltrations in the infundibular region, without significant inflammatory lesions in other brain areas. The infundibular region is outside of the blood-brain barrier and BDV antigens here are directly exposed to contact with immune system cells. Thus, the infundibular region is more susceptible to damage from inflammation than other CNS areas. Damage to the ventral paramedian hypothalamus near the infundibulum leads to the development of an obesity syndrome, as shown by the stereotaxically produced symmetrical lesions (SZENTÁGOTHAI et al. 1972; DUGGAN and BOOTH 1986). Another hypothalamic center related to regulation of body weight is the paraventricular nucleus. Lesions here can also produce experimentally hyperphagia leading to an obesity syndrome (COX and SIMS 1988). In our series, vacuolar degeneration of neurons in this nucleus has frequently been found in the absence of inflammatory changes.

Finally, it is important to consider the role of severe involution of the hippocampus and pyriform cortex in the obesity syndrome. This type of lesion is analogous to bilateral temporal lobectomy, which results in the Klüver-Bucy syndrome (KLÜVER and BUCY 1938, 1939; TERZIAN and ORE 1955). Hyperphagia is a key feature of this syndrome.

It is not possible from current data to determine the significance of pathological changes in the infundibulum, the paraventricular nucleus and the hippocampus in the obesity syndrome. Studies are in progress to review the neuropathology of persistently infected animals that do not have this syndrome.

Virus-induced obesity is not a unique feature of BDV infection; in mice, a similar syndrome was described in the course of canine distemper virus infection (LYONS et al. 1982; NAGASHIMA et al. 1992) and has recently been found also in BDV infected mice (Kao, Gosztonyi and Ludwig 1993, unpublished).

4.7 The Role of Cellular Immunity

Investigators of BD using immunological techniques have been very much impressed by the fact that adoptive transfer of lymphocytes of acutely ill rats to persistently infected rats, first performed by KAO (1985), turns the oligosymptomatic, inapparent infection into an acute, fatal encephalitis. The analogy to lymphocytic choriomeningitis (LILLIE and ARMSTRONG 1945; LEHMANN-GRUBE 1971, 1982; BAENZINGER 1986) was apparent, and on this basis Borna encephalitis was regarded as an immune mediated disease (RICHT 1988; DESCHL 1988; STITZ et al. 1991). The non-lytic nature of tissue culture cells also seemed to corroborate this view. However, at that time the parenchymal damage in various types of BDV infections, as described above, has not been studied with sufficient scrutiny. Persistent infection leads to slowly evolving encephalopathy. On the other hand, infection of newborn rats with a high passage virus leads to a severe, rapidly fatal neurological illness, comparable in every respects to acute BD. Here again, no inflammatory reaction, only a moderate, diffuse infiltration by macrophages could be assessed in animals with relatively longer survival. These observations point to the fact that virus-induced parenchymal damage is mainly responsible for the symptomatology and progression of the disease. It is understandable that the development of a cellular immune reaction may bring about a dramatic deterioration of brain functions. But brain cells are in this constellation passive, "innocent bystanders" of a cellular immune reaction directed against a foreign antigen. If we consider that the majority of neurons, due to persistent virus infection, are functionally more or less compromised, it is conceivable that such an environmental change is deleterious to their function and survival.

A recent study (Petrov 1993) examined the composition and topographical distribution of inflammatory cells in acute experimental Borna encephalitis of the Wistar rat. As referred to above, the overwhelming majority of the immunocompetent T lymphocytes is localized in the leptomeningeal and adventitial spaces, i.e. outside the brain parenchyma. Thus, the conditions of a cell mediated cytotoxicity are not fulfilled because of the lack of contact between T cells and neurons; furthermore, the question of expression of MHC class I gene product on the neuronal cytomembrane is still controversial (Planz et al. 1993; Oldstone and Rall 1993). The cell type which infiltrates the brain parenchyma is the monocyte/macrophage and the intensity of this infiltration is in many cases quite excessive. Recently, much attention has been devoted to the macrophage proliferation in the brain in various virus infections and to the deleterious effects of cytokines and other mediators produced by these cells (Gosztonyi 1992). It is possible that these products are deleterious to the function of neurons, e.g. by disrupting synaptic contacts. These effects may aggravate the latent or manifest virus-induced cytopathic damage.

5 Summary

Natural BD is a nonpurulent acute/subacute encephalitis of horses and sheep with a propensity to involve the olfactory and limbic systems, and the brain stem. The inflammation is concentrated primarily in the gray matter, but subcortical white matter may also be affected. Experimental BD can be produced in a series of animals from birds to primates. The neuropathology after experimental infection is similar to that in natural disease but the inflammatory changes are more diffuse. In the rat and mouse, a persistent/tolerant infection can also be induced, in which inflammatory changes are conspicuously absent. In the course of persistent infection of the rat, an elective, focal degeneration ensues that involves the dentate gyrus, retina, and, less frequently, the magnocellular part of the hippocampus. The cytopathic destruction of the dentate gyrus is the likely anatomical substrate of learning deficiencies and behavioral changes, prominent features of chronic infection. Later in infection, more diffuse and random degeneration of neurons can be found. In all species infected, viral antigens are produced in excess and fill all neuronal processes. Beside neurons, glial cells are infected as well. The agent spreads in the nervous system axonally and transsynaptically (transneuronally). The type of neurotransmitter receptors in the synapse and their interaction with viral proteins may modulate the spread of infection (GOSZTONYI et al. 1994). Virus particles have not been visualized in the brain in any phase of the disease. During persistent infection of the rat, production of viral proteins has a phasic character. Some rats survive acute infection and develop an obesity syndrome. The anatomical basis of this syndrome is not fully clarified; inflammatory destruction of the infundibular region, vacuolar degeneration of the paraventricular nucleus of the hypothalamus and severe, progressive involution of the hippocampal formation most probably play an important role in the production of this neuroendocrine syndrome. In the acute disease, inflammatory reaction can severely aggravate virus-induced cytopathology, but cannot be the sole cause of the neurological disease, since infection with high passage virus can lead to a similarly severe disease in the absence of inflammatory changes.

Acknowledgments: We are grateful to Renate Ehrnsperger and Michaela Schulze for excellent technical assistance. Thanks are due to colleagues, in particular to Moujahed Kao and Ralf Dürrwald, for supply of animal brains with natural and experimental infection, and to Ian Lipkin for supply of the BDV clones for in situ hybridization studies. Constructive discussions with Liv Bode are thankfully acknowledged. This study has been supported by the Deutsche Forschungsgemeinschaft (DFG), by grants to Georg Gosztonyi (No. Go 426/3-1) and to Hanns Ludwig (No. Lu 142/5-1 and 142/5-2), as well as by a grant of the European Union (No. BMH-I-CT 94-1791).

References

Anzil AP, Blinzinger K, Mayr A (1973) Persistent Borna virus infection in adult hamsters. Arch Gesamte Virusforsch 40: 52–57

Avdaloff WM, Gierend M, Sasaki S, Ludwig H (1981) EEG changes in BD virus infected rabbits. Excerpta Med Int Congr Ser 548: 203

Baenzinger J, Hengartner H, Zinkernagel RM, Cole GA (1986) Induction or prevention of immunopathological disease by cloned cytotoxic T cell lines specific for lymphocytic choriomeningitis virus. Eur J Immunol 16: 387–393

Beck A (1925) Die enzootische Encephalitis des Schafes. Dtsch Tierarztl Wochenschr 34: 764

Beck A, Frohböse H (1926) Die enzootische Encephalitis des Schafes. Vergleichende experimentelle Untersuchungen über die seuchenhafte Gehirnrückenmarksenzündung der Pferde und Schafe. Arch Wiss Prakt Tierheilkd 83: 84–110

Bestetti G (1976) Mikroskopische und ultrastrukturelle Untersuchungen an Joest-Degen'schen Einschlusskörperchen bei spontaner Borna-Krankheit des Pferdes. Schweiz Arch Tierheilkd 118: 493–498

Blinzinger K, Anzil AP (1973) Large granular nuclear bodies (karyosphaeridia) in experimental Borna virus infection. J Comp Pathol 83: 589–596

Bohne A (1907) Die Negrischen Körperchen und ihre Bedeutung für die Diagnose der Tollwut. Z Inf Krkh Haust (Berl) 2: 229–242

Briese T, de la Torre JC, Lewis A, Ludwig H, Lipkin WI (1992) Borna disease virus, a negative-strand RNA virus, transscribes in the nucleus of infected cells. Proc Natl Acad Sci USA 89: 11486–11489

Brion JP, Couck AM, Flament-Durant J (1982) Intranuclear inclusions in the neurons of senescent rats Acta Neuropathol (Berl) 58: 107–110

Carbone KM, Duchala CS, Griffin JW, Kincaid AL, Narayan O (1987) Pathogenesis of Borna disease in rats: Evidence that intra-axonal spread is the major route for virus dissemination and the determinant for disease incubation. J Virol 61: 3431–3440

Carbone KM, Trapp BD, Griffin JW, Duchala CS, Narayan O (1989) Astrocytes and Schwann cells are virus-host cells in the nervous system of rats with Borna disease. J Neuropathol Exp Neurol 48: 631–644

Carbone KM, Moench TR, Lipkin WI (1991a) Schwann cells and ependymal cells in persistently infected rats: location of viral genomic and messenger RNAs by in situ hybridization. J Neuropathol Exp Neurol 50: 205–214
Carbone KM, Park SW, Rubin SA, Waltrip RW, Vogelsang GB (1991b) Borna disease: association with a maturation defect in the cellular immune response. J Virol 65: 6154–6164
Cervós-Navarro J, Roggendorf W, Ludwig H, Stitz L (1981) Die Borna-Krankheit beim Affen unter besonderer Berücksichtigung der encephalitischen Reaktion. Verh Dtsch Ges Pathol 65: 208–212
Cox JE, Sims JS (1988) Ventromedial hypothalamic and paraventricular nucleus lesions damage a common system to produce hyperphagia. Behav Brain Res 28: 297–308
Czub S (1988) Das optische System in der Pathogenese der Borna-Krankheit: ein Modell für Virusinfektionen des zentralen Nervensystems. Vet med dissertation, Berlin
De la Torre JC, Carbone KM, Lipkin WI (1990) Molecular characterization of the Borna disease agent. Virology 179: 853–856
Deschl U (1988) Immunhistologische Untersuchungen zur Charakterisierung der Entzündungszellen bei der experimentellen Bornaschen Krankheit der Lewisratte. Vet med dissertation, Giessen
Dittrich W, Bode L, Ludwig H, Kao M, Schneider K (1989) Learning deficiencies in Borna disease virus-infected but clinically healthy rats. Biol Psychiatry 26: 818–828
Duggan JP, Booth DA (1986) Obesity, overeating, and rapid gastric emptying in rats with ventromedial hypothalamic lesions. Science 231: 609–611
Dürrwald R (1993) Die natürliche Borna-Virus-Infektion der Einhufer und Schafe: Untersuchungen zur Epidemiologie, zu neueren diagnostischen Methoden (ELISA, PCR) und zur Antikörperkinetik bei Pferden nach Vakzination mit Lebendimpfstoff. Vet med dissertation, Berlin
Gierend M (1982) Zur Pathogenese der Bornaschen Krankheit: Untersuchungen über die zelluläre Immunantwort, die immunosuppressive Behandlung und die Elektroencephalographie (EEG). Vet med dissertation, Berlin
Gosztonyi G (1978) Axonal and transsynaptic spread of viral nucleocapsids in fixed rabies virus encephalitis. J Neuropathol Exp Neurol 37: 618
Gosztonyi G (1979) Possible mechanisms of spread of fixed rabies virus along neural pathways. In: Bachmann PA (ed) Proceedings of the 4th Munich symposium on mechanisms of viral pathogenesis and virulence. WHO Collaborating Centre for Collection and Evaluation of Data on Comparative Virology, Munich, pp 323–348
Gosztonyi G (1985) Über die elektive Vulnerabilität (spezielle Neurotropie) bei Virusinfektionen des Nervensystems. In: Frydl V (ed) Drittes Neuropathologisches Symposium im Bezirkskrankenhaus Haar. Haar, Germany, pp 69–85
Gosztonyi G (1986) Verbreitung von Viren entlang Neuronennetzen durch transsynaptische Passage–ein Beitrag zur Pathogenese der Tollwut. Tierarztl Prax 14: 199–204
Gosztonyi G (1992) Acute viral encephalitis—a monocyte/macrophage induced disease? Clin Neuropathol 11: 261–262
Gosztonyi G (1994) Ultrastructural composition of lyssaviruses. In: Rupprecht CE et al. (eds) Lyssaviruses. Springer, Berlin Heidelberg New York (Current topics in microbiology and immunology vol 187)
Gosztonyi G, Ludwig H (1984a) Borna disease of horses: an immunohistochemical and virological study of naturally infected animals. Acta Neuropathol (Berl) 64: 213–221
Gosztonyi G, Ludwig H (1984b) Neurotransmitter receptors and viral neurotropism. Neuropsychiatr Clin 3: 107–114
Gosztonyi G, Leiskau T, Ludwig H (1983) The significance of the Borna disease virus infection for the non-mammal, the chicken. Zentralbl Bakteriol Mikrobiol Hyg [A] 225: 170
Gosztonyi G, Kao M, Lefert R, Ludwig H (1984) Propagation of Borna disease virus along neuronal chains: a contribution to the pathogenesis of infection of the central nervous system. Zentralbl Bakteriol Mikrobiol Hyg [A] 258: 510
Gosztonyi G, Kao M, Ludwig H (1985) Neural propagation of Borna disease virus in mice following intranasal inoculation. Zentralbl Bakteriol Mikrobiol Hyg [A] 260: 474
Gosztonyi G, Kao M, Doering-Manteuffel S, Ludwig H (1987) Tissue tropism of Borna disease virus. Zentralbl Bakteriol Mikrobiol Hyg [A] 267: 154
Gosztonyi G, Kao M, Bode L, Ludwig H (1988) Vacuolar myelopathy in rats induced by persistent infection with Borna disease virus. Zentralbl Bakteriol Mikrobiol Hyg [A] 269: 525
Gosztonyi G, Ludwig H, Dietzschold B, Kao M (1990) Transsynaptic virus propagation: an in vivo transfection phenomenon. Acta Neurol Scand 81: 267
Gosztonyi G, Kao M, Bode L, Ludwig H (1991a) Obesity syndrome in experimental infection of rats with Borna disease virus. Clin Neuropathol 10: 33–34

Gosztonyi G, Briese T, Bode L, Lipkin WI, Ludwig H (1991b) Immunohistochemical and molecular detection of Borna disease virus specific structures in the brain. Clin Neuropathol 10: 262

Gosztonyi G, Dietzschold B, Kao M, Rupprecht CE, Ludwig H, Koprowski H (1993) Rabies and Borna disease—a comparative pathogenetic study of two neurovirulent agents. Lab Invest 68: 285–295

Gosztonyi G, Kao M, Dietzschold B, Rupprecht CR, Ludwig H, Koprowski H (1994) Interaction between viral proteins and neurotransmitter receptors directs and modulates transsynaptic virus spread within the central nervous system. Brain Pathol 4: 383

Gosztonyi G, Briese T, Bode L, Lipkin IW, Ludwig H (1995) Ancient (1925) and recent (1991) Borna disease virus "strains" are closely related (in preparation)

Hirano N, Kao M, Ludwig H (1983) Persistent, tolerant or subacute infection in Borna disease virus infected rats. J Gen Virol 64: 1521–1530

Joest E, Degen K (1909) Über eigentümliche Kerneinschlüsse der Ganglienzellen bei der enzootischen Gehirn-Rückenmarksentzündung der Pferde. Z Inf Krkh Haustiere 6: 348–356

Joest E, Degen K (1911) Untersuchungen über die pathologische Histologie, Pathogenese und postmortale Diagnose der seuchenhaften Gehirn-Rückenmarksentzündung (Bornasche Krankheit) des Pferdes. Z Inf Krkh Haustiere 9: 1–98

Joest E, Semmler W (1911) Weitere Untersuchungen über die seuchenhafte Gehirn-Rückenmarksentzündung (Bornasche Krankheit) des Pferdes mit besonderer Berücksichtigung des Infektionsweges und der Kerneinschlüsse. Z Inf Krkh Haustiere 10: 293–320

Kao M (1985) Die Pathogenese der Borna Krankheit bei der Ratte. Ein Modell für persistierende Infektionen und subakute/akute Krankheiten des Zentralnervensystems und für die Fettsucht (Obesity Syndrome). Vet med dissertation, Berlin

Kao M, Gosztonyi G, Ludwig H (1983) Obesity syndrome in Borna disease virus infected rats. Zentralbl Bakteriol Mikrobiol Hyg [A] 255: 173

Kao M, Ludwig H, Gosztonyi G (1984) Adaptation of Borna disease virus to the mouse. J Gen Virol 65: 1845–1849

Kao M, Bode L, Gosztonyi G, Ludwig H (1990) Escape from lethal disease in rats after Borna disease virus infection: survival with obesity syndrome. VIIIth International Congress of Virology, Berlin 1990, abstracts, p 108

Kao M, Gosztonyi G, Bode L, Ludwig H (1995) Pathogenesis of wild and adapted Borna disease (BD) virus strains (in preparation)

Klüver H, Bucy PC (1938) An analysis of certain effects of bilateral temporal lobectomy in the rhesus monkey, with special reference to "psychic blindness". J Psychol 5: 33–54

Klüver H, Bucy PC (1939) Preliminary analysis of functions of the temporal lobes in monkeys. Arch Neurol Psychiatry 42: 979–1000

Környey S (1939) Die primär neurotropen Infektionskrankheiten des Menschen. Fortschr Neurol Psychiat 11: 82–100

Környey S (1954) Das Prinzip der speziellen Neurotropie bei den Viruskrankheiten. Acta Med Acad Sci Hung 6 [Suppl 1]: 119–123

Krey HF, Ludwig H, Boschek CB (1979) Multifocal retinopathy in Borna disease virus infected rabbits. Am J Ophthalmol 87: 157–164

Krey HF, Stitz L, Ludwig H (1982) Virus-induced pigment epithelitis in rhesus monkeys. Clinical and histological findings. Ophthalmologica 185: 205–213

Lehmann-Grube F (1971) Lymphocytic choriomeningitis virus. Springer, Vienna, New York (Virology monographs, vol 10)

Lehmann-Grube F (1982) Lymphocytic choriomeningitis virus. In: Foster HI, Small JD, Fox JG (eds) The mouse in biochemical research, vol 2. Academic, New York, pp 231–266

Lillie RD, Armstrong C (1945) Pathology of lymphocytic choriomeningitis in mice. Arch Pathol 40: 141–152

Lipkin WI, Carbone KM, Wilson MC, Duchala CS, Narayan O, Oldstone BA (1988) Neurotransmitter abnormalities in Borna disease. Brain Res 475: 366–370

Lipkin WI, Travis GH, Carbone KM, Wilson MC (1990) Isolation and characterization of Borna disease agent cDNA clones. Proc Natl Acad Sci USA 87: 4184–4188

Ludwig H, Thein P (1977) Demonstration of specific antibodies in the central nervous system of horses naturally infected with Borna disease virus. Med Microbiol Immunol 163: 215–226

Ludwig H, Becht H, Groh L (1973) Borna disease (BD), a slow virus infection-biological properties of the virus. Med Microbiol Immunol 158: 275–289

Ludwig H, Kraft W, Kao M, Gosztonyi G, Dahme E, Krey H (1985) Borna-Virus-Infection (Borna-Krankheit) bei natürlich und experimentell infizierten Tieren: ihre Bedeutung für Forschung und Praxis. Tierärztl Prax 13: 421–453

Ludwig H, Bode L, Gosztonyi G (1988) Borna disease: a persistent virus infection of the central nervous system. Prog Med Virol 35: 107–151

Ludwig H, Furuya K, Bode L, Klein N, Dürrwald R, Lee DS (1993) Biology and neurobiology of Borna disease viruses (BDV), defined by antibodies, neutralizability and their pathogenic potential. Arch Virol [Suppl] 7: 111–133

Lundgren AL (1992) Feline non-suppurative meningoencephalomyelitis: Clinical and pathological study. J Comp Pathol 107: 411–425

Lundgren AL, Ludwig H (1993) Clinically diseased cats with non-suppurative meningoencephalomyelitis have Borna disease virus-specific antibodies. Acta Vet Scand 34: 101–103

Lundgren A-L, Czech G, Bode L, Ludwig H (1993) Natural Borna disease in domestic animals other than horses and sheep. J Vet Med (B) 40: 298–303

Lyons MJ, Faust IM, Hemmes RB, Buskirk DR, Hirsch J, Zabriskie JB (1982) A virally induced obesity syndrome in mice. Science 216: 82–85

Malkinson M, Weisman Y, Ashash E, Bode L, Ludwig H (1993) Borna disease in ostriches. Vet Record 133: 304

Morales JA, Herzog S, Kompter C, Frese K, Rott R (1988) Axonal transport of Borna disease virus along olfactory pathways in spontaneously and experimentally infected rats. Med Microbiol Immunol 177: 51–68

Müller FL, Fritsch R (1955) Die Augenveränderungen bei der Bornaschen Krankheit. Wien Tierärztl Mschr 42: 866–871

Nagashima K, Zabriskie JB, Lyons MJ (1992) Virus-induced obesity in mice: association with a hypothalamic lesion. J Neuropathol Exp Neurol 51: 101–109

Narayan O, Herzog S, Frese K, Scheefers H, Rott R (1983a) Behavioral disease in rats caused by immunopathological responses to persistent Borna virus in the brain. Science 220: 1401–1403

Narayan O, Herzog S, Frese K, Scheefers H, Rott R (1983b) Pathogenesis of Borna disease in rats: Immune-mediated viral ophthalmoencephalopathy causing blindness and behavioral abnormalities. J Infect Dis 148: 305–315

Neubert M (1984) Die Bornasche Krankheit beim Rhesusaffen—eine licht- und elektronenmikroskopische Untersuchung. Inaugural-dissertation, Berlin 1984

Nitzschke E (1963) Untersuchungen über die experimentelle Bornavirus-Infektion bei der Ratte. Zentralbl Vet Med (B) 10: 470–527

Oldstone MBA, Rall GF (1993) Mechanism and consequence of viral persistence in cells of the immune system and neurons. Intervirology 35: 116–121

Orioli PJ, Strick PL (1989) Cerebellar connections with the motor cortex and the arcuate premotor area: An analysis employing retrograde transneuronal transfer of WGA-HRP. J Comp Neurol 288: 612–626

Petrov (1993) Analyse der entzündlichen Reaktion bei der experimentellen Borna-Encephalitis der Ratte. Inaugural dissertation, Freie Universität Berlin

Pette H, Környey S (1935) Über die Pathogenese und die Pathologie der Bornaschen Krankheit im Tierexperiment. Dtsch Z Nervenheilkd 136: 20–65

Planz O, Bilzer T, Sobbe M, Stitz L (1993) Lysis of major histocompatibility complex class I-bearing cells in Borna disease virus-induced degenerative encephalopathy. J Exp Med 178: 163–174

Richt J (1988) Bedeutung der zellulären Immunreaktion bei der Pathogenese der Bornaschen Krankheit. Vet med dissertation, Giessen

Roggendorf W, Sasaki S, Ludwig H (1983) Light microscope and immunohistological investigations on the brain of Borna disease virus-infected rabbits. Neuropathol Appl Neurobiol 9: 287–296

Rubin SA, Waltrip RW, Bautista JR, Carbone KM (1993) Borna disease virus in mice: Host-specific differences in disease expression. J Virol 67: 548–552

Schmidt J (1912) Untersuchungen über das klinische Verhalten der seuchenhaften Gehirn-Rückenmarksentzündung (Bornaschen Krankheit) des Pferdes nebst Angaben über diesbezügliche therapeutische Versuche. Berl Tierärztl Wochenschr 28: 581–586 and 597–603

Schneider P, Briese T, Zimmermann W, Ludwig H, Lipkin WI (1994) Sequence conservation in field and experimental isolates of Borna disease virus. J Virol 68: 63–68

Seifried O (1931) Pathologie neurotroper Viruskrankheiten der Haustiere (mit Berücksichtigung der vergleichenden Pathologie). Erg Allg Pathol 24: 554–676

Seifried O, Spatz H (1930) Die Ausbreitung der encephalitischen Reaktion bei der Bornaschen Krankheit der Pferde und deren Beziehungen zu der Encephalitis epidemica, der Heine-Medinschen Krankheit und der Lyssa des Menschen. Eine vergleichend-pathologische Studie. Z Neurol Psychiat 124: 317–382

Shadduck JA, Danner K, Dahme E (1970) Fluoreszenzserologische Untersuchungen über Auftreten und

Lokalisation von Borna-Virusantigen in Gehirnen experimentell infizierter Kaninchen. Zentralbl Vet Med (B) 17: 453–459
Sprankel H, Richarz K, Ludwig H, Rott R (1978) Behavior alterations in tree shrews (Tupaia glis, Diard 1820) induced by Borna disease virus. Med Microbiol Immunol 165: 1–18
Stitz L, Richt JA, Rott R (1991) Immunpathogenese der Borna-Krankheit. Tierarztl Prax 19: 267–270
Szentágothai J, Flerkó B, Mess B, Halász B (1972) Hypothalamic control of the anterior pituitary. An experimental-morphological study, 3rd edn. Akadémiai Kiadó, Budapest
Terzian H, Ore GD (1955) Syndrome of Klüver and Bucy—reproduced in a man by bilateral removal of the temporal lobes. Neurology 5: 373–380
Vogt C, Vogt O (1920) Zur Lehre der Erkrankungen des striären Systems. J Psychol Neurol (Lpz) Ergebn-Heft 3, 25: 627
Wagner K, Ludwig H, Paulsen J (1968) Fluoreszenzserologischer Nachweis von Borna-Virus Antigen. Berl Münch Tierärztl Wochenschr 81: 395–396
Walther A (1952) Auffällige Sehstörungen bei Bornascher Krankheit des Pferdes. Dtsch Tierärztl Wochenschr 59: 88
Weisman Y, Malkinson M, Perl S, Machany S, Lublin A, Ashash E (1993a) Paresis in young ostriches. Vet Rec 133: 78
Weisman Y, Malkinson M, Ashash E, Nir A (1993b) Serum therapy of a paretic syndrome of ostriches. Vet Rec 133: 172
Zimmermann W, Dürrwald R, Ludwig H (1993) Detection of Borna disease virus RNA in naturally infected animals by the polymerase chain reaction. J Virol Methods 46: 133–143
Zwick W (1939) Bornasche Krankheit und Encephalomyelitis der Tiere. In: Gildenmeister E, Haagen E, Waldmann O (eds) Handbuch der Viruskrankheiten II. Fischer, Jena, pp 254–354
Zwick W, Seifried O (1927) Infektiöse Gehirn-Rückenmarks-Entzündung (Bornasche Krankheit) des Pferdes. In: Kolle W, Kraus R, Uhlenhuth P (eds) Handbuch der pathogenen Mikroorganismen. Fischer, Jena, Urban and Schwarzenberg, Berlin, pp 117–144
Zwick W, Seifried O, Witte J (1926) Experimentelle Untersuchungen über die seuchenhafte Gehirn- und Rückenmarksentzündung der Pferde (Bornasche Krankheit). Z Inf Krkh Haustiere 30: 42–136
Zwick W, Seifried O, Witte J (1929) Weitere Beiträge zur Erforschung der Bornaschen Krankheit des Pferdes. Arch Wiss Prakt Tierheilkd 59: 511–545

Note Added in Proof

The Borna disease virus (BDV) or structural subunits of the virion have not clearly been identified in naturally or experimentally infected animals. During the publishing process of this book we succeeded, however, to characterize cell-free virus released from infected tissue cultures and demonstrate its specificity by immunoelectron microscopy (Zimmermann W, Breter H, Rudolph M, Ludwig H (1994) J Virol 68: 6755–6758).

Immunopathogenesis of Borna Disease

L. STITZ[1], B. DIETZSCHOLD[2], and K.M. CARBONE[3]

1 Introduction . . . 75

2 General Features of Borna Disease and Borna Disease Virus . . . 77

3 Experimental Borna Disease in Rats . . . 77
3.1 Borna Disease Virus Infection in Immunocompetent Rats . . . 78
3.2 Borna Disease Virus Infection in Immunoincompetent Rats . . . 79
3.2.1 Aspects of Borna Disease Virus Infection in Newborn and Athymic Rats . . . 79
3.2.2 Aspects of Borna Disease Virus Infection in Immunocompromised Rats . . . 80

4 Cellular Immune Response in Borna Disease . . . 82
4.1 Characterization of Inflammatory Cells and MHC-Expressing Cells Involved in the Immunopathological Reaction . . . 82
4.2 Pathogenic Relevance of Borna Disease Virus-Specific CD4+ T Cells . . . 82
4.3 Pathogenic Relevance of CD8+ T Cells in Borna Disease . . . 84
4.3.1 Influence of T Cell-Specific Monoclonal Antibodies In Vivo . . . 84
4.3.2 Influence of Cytokine Treatment on the Immunopathogenesis of Borna Disease . . . 84
4.3.3 Presence of Cytotoxic T Lymphocytes in the Brain of Borna Disease Virus and Their Relative Importance in Brain Atrophy . . . 85

5 Cytokine Expression in the Central Nervous System of Borna Disease Virus-Infected Rats . . . 87

6 Conclusion . . . 88

References . . . 89

1 Introduction

Diseases of the central nervous system are fearsome conditions due to their deleterious effects on physical and mental functions. In addition to acute diseases caused by viruses and bacteria infecting the meninges and the brain, disturbances of motility, disorders in sensory functions, behavioral abnormalities, personality changes and chronic debility and dementia can be long-lasting consequences after infection of the central nervous system (CNS) by well known

[1] Institut für Virologie, Justus-Liebig-Universität, Frankfurter Strasse 107, 35392 Gießen, Germany
[2] Thomas Jefferson University, Center for Neurovirology, Dept. of Microbiology and Immunology, 1020 Locust Street, Philadelphia, PA 19107, USA
[3] The Johns Hopkins University, School of Medicine, Division of Infectious Diseases, Ross 1159, Baltimore, MD 21205, USA

microbes. Furthermore, yet unknown mechanisms or uncharacterized agents may induce CNS diseases. Pathological alterations linked to viruses or virus-like agents, such as the demonstration of HIV-1 antigen in the brains of patients suffering from dementia, have stimulated great public interest in this field (PRICE and BREW 1988; PRICE et al. 1988). This concern is fueled further by the recent appearance of "mad-cow disease" in milk and meat producing cattle caused by an unidentified "Scrapie agent" (HOPE et al. 1988). The recent discussions on whether the latter disease can be transmitted to humans by consumption of food products from infected cattle (WILESMITH et al. 1988) are validated by the demonstration of transmission of the human spongiform encephalopathy, Creutzfeldt-Jakob-Scheinker syndrome or Kuru-Kuru, after transplantation of tissue from infected donors or by cannibalism of afflicted human victims of this syndrome (GAJDUSEK and GIBBS 1971; DUFFY et al. 1974; MANUELIDIS and MANUELIDIS 1988).

Apart from these afflictions of the CNS in which no immunological response has yet been determined, there is increasing evidence that immunological mechanisms are involved in a considerable number of infectious disease processes of the CNS. Many autoimmune and virus-induced immunopathological alterations of the brain have been studied in experimental animals and serve as models for animal and human diseases. In infection of mice with the lymphocytic choriomeningitis virus, or Theiler's virus, the presence and pathogenic importance of a virus-specific immune response, especially mediated by T cells, has been demonstrated in the brain (COLE et al. 1972; DOHERTY and ZINKERNAGEL 1974; YAMADA et al. 1991). Other rodent models studied include corona virus and measles virus infections of rats, in which an autoimmune reaction is of crucial importance in the pathogenesis of virus-induced diseases of the brain (NAGASHIMA et al. 1978; LIEBERT et al. 1988). In addition to HIV in AIDS encephalopathy, several observations in human patients support this view, namely the detection of measles virus in cases of subacute sclerosing panencephalopathy (STEPHENSON and TER MEULEN 1979), and of papovavirus in progressive multifocal leukoencephalopathy (PADGETT et al. 1971; WEIMER et al. 1972). Furthermore, several viruses have been suggested to be involved in the development of multiple sclerosis (TER MEULEN and STEPHENSON 1983)

In general, the outcome of a viral infection depends greatly on the efficiency and the speed of the immune system's reaction to the invading agent, in addition to characteristics of the virus. Viruses that are highly cytopathic represent a more immediate threat to the life of their hosts than non- or poorly cytopathic viruses. Therefore, it is not surprising that the immune system is designed for efficient and rapid elimination of the cytopathic viruses in order to avoid spread of the infection and to reduce tissue destruction. However, a vigorous, destructive immune reaction might have considerable deleterious effects on the host in the case of infection with a persistent noncytopathic virus. Since the immune system is apparently not capable of distinguishing between cytolytic and noncytolytic viruses, the resulting damage to the host by the immune response to a nonlytic agent might follow infection with an agent which is otherwise relatively innocuous.

2 General Features of Borna Disease and Borna Disease Virus

An example of such a noncytopathic virus is Borna disease virus (BDV). Until recently BDV was thought to be a natural pathogen only in horses and sheep in certain areas of Europe. However, serological data suggest that Borna disease (BD) may be more widely distributed geographically and may have a more extensive host range including humans suffering from psychiatric disorders (STITZ and ROTT 1993; CAPLAZI et al. 1994; see chapter by Rott and chapter by Bode). Therefore BDV has to be regarded as a true zoonotic agent (STITZ et al. 1993). The virus and the disease have been named after the town of Borna in Saxony, where, in 1895, an epidemic among horses of a cavalry regiment resulted in the loss of a large number of animals. At that time, the disease had already been recognized for about 100 years but was known under various synonyms reflecting CNS disorders such as "disease of the head" and "infectious brain and spinal cord inflammation." A particular characteristic of the disease is the invariably long incubation period between exposure or inoculation and development of disease, resulting in the grouping of BDV with other "slow virus" infections. In the experimental rat model, some variability in incubation period has been linked to the route of inoculation (CARBONE et al. 1987). The disease is associated with disturbances of motility and in sensory functions and usually results in paralysis and death in affected natural hosts (reviewed in chapter by Rott). Pathohistologically, BD is classified as a progressive polioencephalomyelitis with pronounced mononuclear inflammatory reactions in the basal cortex, the nucleus caudatus and the entire hippocampal area (reviewed in chapter by Gosztonyi). Recently, evidence of infection and inflammatory response in the peripheral and autonomic nervous system has also been demonstrated (CARBONE et al. 1987, 1989).

BDV has been only recently characterized as a negative sense single-stranded RNA virus (reviewed in chapter by Briese). The virus, which is tightly cell-associated, lacks apparent cytopathogenicity in vitro (HERZOG and ROTT 1980). In vivo, the virus replicates preferentially in cells derived from the neural crest such as neurons, astrocytes and ependymal cells (NARAYAN et al. 1983a; CARBONE et al. 1989, 1991a; DESCHL et al. 1990). In addition, after cocultivation (HERZOG and ROTT 1980) or after repeated infection with clarified homogenates from infectious rat brain, other cell types such as skin, kidney, testis, spleen cells and astrocytes (DANNER et al. 1978; HERZOG and ROTT 1980; RICHT and STITZ 1992; PLANZ et al. 1993) can be directly infected in vitro.

3 Experimental Borna Disease in Rats

After experimental infection, a wide variety of vertebrates can be infected, including species phylogenetically distant, ranging from birds to monkeys. The species that has been studied most intensively is the rat and most, if not all, progress in understanding the pathogenesis of this virus-induced disease of the CNS was achieved by using this experimental animal.

3.1 Borna Disease Virus Infection in Immunocompetent Rats

After intracerebral (i.c.) or intranasal (i.n.) infection of adult rats, infectious virus and virus-specific antigen can be detected in high concentrations in the brain, retina, cerebrospinal fluid, peripheral nerves and adrenal gland (NARAYAN et al. 1983a,b; CARBONE et al. 1987, 1989; DESCHL et al. 1990). Antigen can be found in the brain in ependymal cells as early as 4 days after infection. Thereafter, BDV-specific antigen can be also detected in the nucleus and cytoplasm of neurons, astrocytes and oligodendrocytes in all cortical and brain stem areas. Endothelial cells have not been reported to contain virus-specific antigen. Furthermore, in acute BD, virus or virus-specific antigen has not been detected in inflammatory cells. In the eye, antigen can be found in all retinal layers. At later stages of the infection the virus disappears from the retina and the animals become blind, but a persistent, productive infection is maintained in the other tissues mentioned above (NARAYAN et al. 1983a). In those animals who survive the acute infection and develop chronic disease, virus antigen can be found in cells of the peripheral nervous system such as Schwann cells (CARBONE et al. 1989), and virus can be located by culture and/or PCR in extraneural tissues (SHANKAR et al. 1992) including peripheral blood mononuclear cells (SIERRA-HONIGMANN et al. 1993). In immunocompetent infected rats, disease signs such as lack of grooming, ataxia, hyperactivity and aggressiveness can be seen at about day 14. This acute phase of the disease lasts for about 3 weeks and surviving animals go on to develop chronic BD. The clinical onset of acute BD is paralleled by the development of a mononuclear inflammatory reaction in the central and peripheral nervous system that is primarily perivascular. However, encephalitic lesions are also found in brain parenchyma. In general, the inflammation is initially centered in the limbic system but spreads to other areas of the brain during infection (NARAYAN et al. 1983a; CARBONE et al. 1987; DESCHL et al. 1990). The cells present in perivascular cuffs have been intensively studied for their phenotypes. It is important to note that the development of neurological symptoms is directly correlated with the appearance of inflammatory cells. T cells appear very early after infection in the brain and macrophages represent the most frequent cell type present in the inflammatory lesions throughout the course of the infection. Additionally, numerous B cells are found, showing maximal numbers later than all other cells participating in the inflammatory reaction.

The chronic phase of BD is clinically characterized by increasing apathy, somnolence, signs of dementia and behavioral abnormalities (see chapter by Solbrig). Interestingly, late, i.e., more than 2–3 months, after infection, the inflammatory reaction starts to decrease, resulting finally in an almost complete absence of inflammation. Brains show dramatic loss of neural tissue and a severe hydrocephalus ex vacuo can be observed.

Some chronically infected rats develop an impressive obesity, with body weights of up to 500–600 g with no or minimal signs of classical BD (see chapter by Rott and Becht).

3.2 Borna Disease Virus Infection in Immunoincompetent Rats

3.2.1 Aspects of Borna Disease Virus Infection in Newborn and Athymic Rats

A different clinical picture is seen in rats that are infected with BDV when they are in a natural or drug-induced state of immunoincompetence. Rats are born at a relatively early stage of development with an imperfectly functioning immune system. In contrast to the enhanced virulence of many cytolytic viruses in newborn rats, infection with BDV within 1 or 2 days after birth produces a persistent infection with few of the clinical hallmarks of classical Borna disease (persistent, tolerant infection of the newborn, PTI-NB). At first glance, PTI-NB rats appear normal, but, upon closer inspection, these rats have clear physical and behavioral abnormalities. For example, PTI-NB rats are smaller in weight and length than age-matched, uninfected rats (Carbone et al. 1991b; Bautista et al. 1994). The slowing of weight gain occurs around the time of weaning, suggesting that PTI-NB rats have abnormal self-feeding behaviors following apparent adequate nursing behavior (Bautista et al. 1994). Unlike runted mice persistently infected with lymphocytic choriomeningitis virus, PTI-NB rats have normal serum levels of growth hormone and glucose (de la Torre and Carbone, unpublished data). In addition, these rats have abnormalities of ingestive behaviors, since PTI-NB rats choose saline over sweet solutions in taste preference tests (Bautista et al. 1994). Other behavioral aberrations have been described, including abnormalities in spatial discrimination learning, increased motor activity, and decreased aversion learning behavior (Dittrich et al. 1989). In addition, a transient increase in diurnal activity can be detected in PTI-NB rats, whereas these animals maintain an abnormally high level of hyperactivity for months after infection (Bautista et al. 1994).

Pathological changes in the CNS are also found in these rats, such as hypoplastic and disorganized cerebella. Unexpectedly, direct viral lysis of certain types of neurons (e.g., dentate gyrus) was suggested by the gradual loss of these cells in PTI-NB rats, even in the absence of encephalitis (Narayan et al. 1983a; Carbone et al. 1991b). Taken as a whole, the abnormalities identified in PTI-NB rats provide evidence for direct viral effects on cell and organism function and offer the opportunity to study viral-induced pathology in the absence of immuopathological sequelae.

The mechanism for immunologic nonresponsiveness to BDV of the PTI-NB rats has not been conclusively demonstrated. However, preliminary characterization of the immunological abnormalities in these rat has produced some intriguing findings. Humoral response to BDV is greatly reduced, as compared to rats infected with BDV as adults, particularly in the first few months after infection

(NARAYAN et al. 1983a; HERZOG et al. 1985; CARBONE et al. 1991b). Adoptive transfer of spleen cells from rats in the acute stage of BD failed to induce encephalitis or disease in PTI-NB recipient rats. In contrast, the BDV-specific immunological non-responsiveness of adult rats infected after the administration of pharmacological immunosuppressants, such as cyclophosphamide or cyclosporine A, is reversed by inoculation with an identical aliquot of spleen cells (NARAYAN et al. 1983a; HIRANO et al. 1983; STITZ et al. 1989; see following section). In experiments using parabiosis and bone marrow transplantation, the defect in immune responses to BDV in PTI-NB rats was shown to be linked to events that occurred during maturation of the immune cells, after exiting the bone marrow (CARBONE et al. 1991b). Intriguingly, it has now been demonstrated that the peripheral blood mononuclear cells and the bone marrow cells from PTI-NB rats are infected with BDV, although the relationship of this finding to the lack of immune response to BDV in these rats is not known (CARBONE et al. 1991b; SIERRA-HONIGMANN et al. 1993). Spread of BDV to other extraneural tissues is also seen in PTI-NB rats (NARAYAN et al. 1983a; HERZOG et al. 1985). These data, combined with similar findings in rats immunosuppressed with cyclosporine A prior to BDV infection (STITZ et al. 1991a; Bilzer et al., unpublished observation), suggest that the absence of a BDV-specific immune response leads to a extraneural dissemination of the virus. However, the observation that the virus is restricted to the nervous system in cyclophosphamide-immunosuppressed rats indicates that the competence of the immune system is not the only factor determining tropism (HERZOG et al. 1984; Bilzer et al., unpublished observation). Even after the neonatal period the age of the animal at the time of infection is important: athymic rats infected at an age of 4 weeks have viral antigen in nonneural tissues; in contrast, nude rats infected at 5 months show restriction of the virus to the CNS (HERZOG et al. 1984). It has therefore been suggested that different patterns of virus dissemination in BDV-infected rats might depend on the stage of maturation and possibly on changes in cellular structures involved in virus infection (HERZOG et al. 1984; Bilzer et al., unpublished observation).

3.2.2 Aspects of Borna Disease Virus Infection in Immunocompromised Rats

Like PTI-NB and athymic rats, cyclophosphamide- or cyclosporine A (CSA)-treated rats infected with BDV also do not show BD or the acute inflammatory reaction. The rationale for using immunocompromised experimental animals came from earlier experiments in which rhesus monkeys had been used (STITZ et al. 1980). It was then observed that monkeys which had undergone splenectomy prior to BDV-infection showed a different clinical and histopathological picture. In contrast to animals not undergoing surgery, which developed severe clinical disease, a marked lymphohistiomonocytic encephalitis and retinitis, splenectomized monkeys showed minimal neurological disturbances, no motor paralysis and fewer cellular infiltrates in the brain, the choroid and the retina (CERVÓS-NAVARRO et al. 1981). In addition, it appeared from in vitro cytotoxicity assays that lymphocytes isolated from splenectomized animals had lower cytotoxic activity (Stitz et al. 1980). Therefore, it was suggested that the cell-mediated immune status might be of importance in BD (Stitz et al. 1980). In a more detailed study, Narayan et al. established the immunopathological basis of Borna disease in rats (Narayan et al. 1983a). In this investigation it was shown that immunosuppression with cyclophosphamide of BDV-infected rats resulted in the absence of encephalitis and disease, whereas these persistently infected tolerant rats were susceptible to the disease after adoptive transfer of immune lymphocytes.

Virus-specific nucleic acid and infectious virus, in addition to virus-specific antigen, are found in immunocompromised animals in amounts comparable to fully immunocompetent rats (Narayan et al. 1983a; Herzog et al. 1985; Carbone et al. 1987; Stitz et al. 1991a). Strikingly, immunocompromised rats, particularly newborn infected animals, show no destruction of the retina, although the virus persists in the eye, and the rats do not become blind despite the presence of virus in retinal layers (Narayan et al. 1983a). Thus, with few exceptions, such as those described in PTI-NB rats, BDV has little to no direct cytopathogenicity in vivo. The importance of the immune response to the pathogenesis of BD was further stressed by showing that adoptive transfer of lymphocytes from BDV-immune rats into immunoincompetent animals resulted in full-blown BD (Narayan et al. 1983a; Stitz et al. 1989). In CSA-immunosuppressed rats it was demonstrated by various approaches that no virus-specific T cells were present. This represented the first model for long-lasting inhibition of an immune-mediated disease in a persistent virus infection. Furthermore, rats were challenged intracerebrally either during the period of CSA treatment or thereafter. In agreement with the inability of a short-term treatment with CSA to inhibit the disease (Stitz et al. 1989), i.c. reinfection before the end of the treatment period led to clinical symptoms and encephalitic lesions whereas reinfection after the discontinuation of CSA did not. These experiments revealed strong evidence that T cell tolerance is induced under CSA-treatment in BDV-infected rats and that the persistence of foreign antigen is the prerequisite and basis for obtaining a tolerant state (Stitz 1992).

Several lines of evidence indicate that antiviral antibodies do not play a significant role in the pathogenesis of BD (Narayan et al. 1983a; Herzog et al. 1985). The most decisive argument against the involvement of antibodies in the pathogenesis of BD comes from experiments with the immunosuppressive drug CSA. Rats treated with CSA can be protected from BD and do not mount an antibody response (Stitz et al. 1989). Interestingly, treatment of BDV infected rats with lower doses of CSA results in the development of an encephalitis in the absence of an anti-BDV antibody response. Additionally, while an i.c. challenge of CSA-treated rats with BDV does not result in immunopathological disease, it does restore immunoreactivity at the B cell level (Stitz et al. 1989). As a whole, these facts together demonstrate that BD is likely due to a virus-induced immunopathological reaction at the T cell level.

4 Cellular Immune Response in Borna Disease

4.1 Characterization of Inflammatory Cells and MHC-Expressing Cells Involved in the Immunopathological Reaction

Although the above mentioned findings clearly indicate that the pathogenesis of BD is closely related to the cellular immune response, they do not reveal the cellular basis of the immunopathological process resulting in inflammation of the brain. By characterizing the cells present in the inflammatory lesions, employing immunohistochemical methods, a useful approach was found to solve this problem. Immunohistological investigations into the quality of cells involved in the perivascular inflammatory reaction, employing monoclonal antibodies, revealed the presence of CD4+ and CD8+ T cells in addition to numerous macrophages and B cells (DESCHL et al. 1990).

Since restriction elements are most important in T cell-mediated immune phenomena, the presence of MHC class I and II antigen was scrutinized in the brains of BDV-infected rats. This self-antigen was detected on various cell types perivascularly but was also on oligodendrocytes, microglial and ependymal cells (DESCHL et al. 1990; RICHT et al. 1990; STITZ et al. 1991b; CARBONE et al. 1991a). It is of note that MHC class II antigen was also detected in areas without inflammation, arguing for a general induction and expression of MHC class II in the brain of BDV-infected rats.

By characterizing the brain cells that express MHC class I antigen it became evident that this self-antigen could be demonstrated on neurons and astrocytes in vivo (BILZER and STITZ 1994) and in vitro (PLANZ et al. 1993).

The expression of MHC antigens alone cannot be regarded as an indicator of an ongoing MHC restricted immune response because enhanced expression of this self-antigen has been found in a variety of chronic infectious and non-infectious encephalitis. Although under normal conditions, the CNS has low levels of MHC antigen, greater expression occurs during pathological situations (VITETTA and CAPRA 1978; LAMPSON 1987).

4.2 Pathogenic Relevance of Borna Disease Virus-Specific CD4+ T Cells

To elucidate the importance of T cell subsets in the pathogenesis of BD in a first approach, a homogeneous virus-specific T cell line was established (RICHT et al. 1989). Lymphocytes obtained from regional lymph nodes after subcutaneous immunization with purified virus-specific antigen were cultured and restimulated in vitro employing a protocol for the cultivation of CD4+ T cells (RICHT et al. 1990). Analysis of this cell line revealed BDV specificity, MHC class II restriction and the phenotypical markers of CD4+ helper/inflammatory cells (RICHT et al. 1989).

Adoptive transfer of this cell line into BDV-infected immunosuppressed healthy recipients resulted in severe disease and death as early as day 5 after the injection of effector cells (RICHT et al. 1989, 1990). In contrast, passive transfer into uninfected rats did not result in encephalitis or disease, demonstrating that this BDV-specific T cell line, by itself, is not encephalitogenic. These results, together with the immunohistological characterization of inflammatory cells in the brain of BDV-infected rats, suggest that a delayed type hypersensitivity reaction (DTH) may play a role in BD. Interestingly, this T cell line has also been shown to induce prevention or enhancement of the immunopathological disease dependent on whether the adoptive transfer occurred before or after infection (RICHT et al. 1994). A similar mechanism has only been observed before as mediated by CD8+ T cells in virus-infected mice (BAENZIGER et al. 1986). Whether this phenomenon in BDV-infected rats is due to CD4+ or CD8+ T cells cannot yet be answered. Unlike corona virus- and measles virus-induced encephalomyelitis, which represent models for virus-induced autoimmune diseases (JAHNKE et al. 1985; LIEBERT et al. 1988) and in which CD4+ cells also have been shown to be of immunopathological relevance, BD is an example of immunopathology in the brain resulting from an immune reaction to viral antigen(s).

The importance of MHC class II-bearing cells in the pathogenic mechanism has not yet been determined, especially since it has not yet been possible to demonstrate their role in antigen presentation of BDV-specific antigen in the brain. However, some evidence has accumulated that provides better insight into the mechanisms of pathogenicity and the T cell subsets involved. In order to define a cell type in the CNS that might be relevant to the in vivo situation in BD, functional interactions between CD4+ BDV-specific T cells and astrocytes were tested in vitro. Astrocytes were selected for two reasons. First, astrocytes have been shown to be target cells of BDV infection in the rat brain (LUDWIG et al. 1988; CARBONE et al. 1989, 1991a; DESCHL et al. 1990). Second, astrocytes are potent antigen presenting cells, suggesting an involvement in immunopathology after viral infections or in autoimmune disorders in the CNS (FONTANA et al. 1984; FIERZ et al. 1985).

We found that induction of MHC class II expression on astrocytes by interferon-γ (IFN-γ) increased the proliferative capacity of the BDV-specific CD4+ T cell line in vitro (RICHT et al. 1990). In contrast to coronavirus infections (MASSA et al. 1986; SUZUMURA et al. 1986), BDV infection alone did not induce MHC class II expression in vitro (RICHT et al. 1990; RICHT and STITZ 1992; PLANZ et al. 1993) or in vivo (STITZ et al. 1991a).

Interestingly, and relevant to the in vivo situation, BDV-infected astrocytes retained their full ability to present BDV-specific antigen (RICHT and STITZ 1992). Furthermore, the expression of MHC class II on BDV-infected astrocytes was a prerequisite to their function as target cells for in vitro cytotoxicity by the CD4+ T cell line (RICHT and STITZ 1992). This BDV-specific CD4+ T cell was able to lyse syngeneic, IFN-γ treated, persistently infected astrocytes and lysis was significantly reduced by antibodies directed against MHC class II antigens. However, it cannot be excluded that this BDV-specific CD4+ T cell line acquired its cytotoxic activity upon in vitro cultivation as reported previously for CD4+ T cells (FLEISCHER 1984). A CD4+ T cell line characterized as a functional T helper cell and specific for another major BDV-coded protein did not exert cytotoxic activity (PLANZ et al. 1995). These in vivo and in vitro findings indicate the importance of virus-specific CD4+ T cells and suggest that lymphokines may be of crucial importance for the induction and maintenance of the immunopathological reaction resulting in BD.

4.3 Pathogenic Relevance of CD8+ T Cells in Borna Disease

4.3.1 Influence of T Cell-Specific Monoclonal Antibodies In Vivo

Due to difficulties in culturing rat CD8+ T cells, investigation into the pathogenic importance of CD8+ T cells in BD has been pursued through experiments in which BDV-infected rats were treated with monoclonal antibodies directed against various T cell markers (STITZ et al. 1992). Antibodies specific for CD4+ and CD8+ cells were able to decrease or even prevent the local inflammatory reaction if given early during the infection. Using a mouse monoclonal antibody directed against a pan-T cell marker (Ox–52), it could be demonstrated that the immunopathological reaction in the brain, as well as the disease, could be completely inhibited in BDV-infected thymectomized rats. Treatment started after infection resulted in delay of the inflammatory reaction, but ultimately did not affect the severity of encephalitis and disease. The limited effect of monoclonals given late after infection indicates that T cells are rapidly activated in the periphery. This may reflect the fact that i.c. infection results in systemic dissemination of the virus (MIMS 1960). Though systemic dissemination is not likely after i.n. experimental infection, this route also resulted in efficient antigen presentation presumably because dendritic cells and macrophages are located at the site of inoculation.

CD8-specific monoclonal antibodies were more effective in treating the immunopathological disease than antibodies directed against CD4+ cells. These experiments indicate that CD8+ T cells play an important role in the pathogenesis of BD. Since CD8+ T cells can produce cytokines, it has been proposed that an initial CD8+ T cell response may be decisive in triggering the local delayed type hypersensitivity reaction in the brain after BDV infection (STITZ et al. 1992). These findings also support the notion that CD8+ T cells are an integral component of a delayed type hypersensitivity reaction.

4.3.2 Influence of Cytokine Treatment on the Immunopathogenesis of Borna Disease

The importance of lymphokines in BD was first shown directly in experiments employing transforming growth factor (TGF)-β2 in vivo (STITZ et al. 1991b). TGF belongs to a class of polypeptides exhibiting diverse effects on cell growth and differentiation. These substances act as multifunctional cytokines with potent inhibitory activity on growth, differentiation and effector functions of both activated T and B lymphocytes and macrophages (reviewed in WAHL et al. 1989; PALLADINO et al. 1990). Experiments with TGF-β2 in BDV-infected rats revealed a reduction of the severity of clinical symptoms that was paralleled by a significant reduction of the inflammatory reaction in the brain. Thus, TGF-β2 delayed the development of T cell-mediated disease and tissue destruction if given systemically (FONTANA et al. 1989; STITZ et al. 1991b). Immunohistological investigations revealed only slightly reduced CD4+ T cell numbers and no changes in macrophage counts in encephalitic lesions of TGF-treated rats whereas CD8+ T cells were virtually absent from the perivascular inflammatory reaction. The expression of MHC class II antigen was significantly reduced in the brain of TGF-treated BDV infected rats, whereas MHC class I expression was not. Since CD8+ T cells, are potent producers of immune IFN, and since IFN-γ also regulates MHC class I and class II expression (WONG et al. 1984; FIERZ et al. 1985), the absence of CD8+ T cells in the brain of TGF-treated rats might be responsible for the observed reduction of MHC class II antigen (STITZ et al. 1991b). The recent finding that BDV-infected astrocytes produce in vitro IFN of the α/β type that shows all characteristics of a previously described astrocyte IFN (TEDESCHI et al. 1986) supports this interpretation (PLANZ et al. 1993). IFN α/β and, in particular, IFN produced by astrocytes, up-regulates MHC class I but not class II expression (TEDESCHI et al. 1986; HALLORAN et al. 1989), which would, in turn, explain the IFN-γ independent presence of MHC class I in TGF-treated rats. The experiments performed with TGF-β2 revealed an initial dramatic reduction in the local immune response after BDV infection. The relative absence of CD8+ T cells seemed to be decisive, since the increase of CD8+ T cells late after TGF treatment was directly correlated with an increase in the expression of MHC class II antigen in the brain and encephalitic lesions. In summary, in spite of the presence of CD4+ T cells, reduced expression of restriction elements for cell-mediated immune response led to an initial inhibition of the encephalitic reaction and clinical disease.

4.3.3 Presence of Cytotoxic T Lymphocytes in the Brain of Borna Disease Virus and Their Relative Importance in Brain Atrophy

The mechanisms by which virus infections of the CNS cause neuronal damage are not fully known. Studies indicate that in many cases the virus does not directly destroy neurons, but may cause indirect damage by triggering cell-mediated immune response and or altering neuron-specific functions. The presence of inflammatory components, notably T cells and macrophages, frequently characterize neurological disease caused by infection with conventional viruses, such as measles (WOLINSKI 1990), rubella (WOLINSKI 1990), HIV (GRANT and ATKINSON 1990; PRICE et al. 1990), lymphocytic choriomeningitis (DOHERTY et al. 1976; BYRNE and OLDSTONE 1984), and also BDV (NARAYAN et al. 1983a; DESCHL et al. 1990). Immunopathology may be mediated by cytokines, neurotoxins and radicals and cytotoxic T lymphocytes (CTLs).

The chronic phase of BD is characterized by a prominent cortical atrophy and hydrocephalus internus ex vacuo (see chapter by GOSZTONYI). This hydrocephalus does not result secondarily from hydrocephalus occlusus (IRIGOIN et al. 1990) nor from hypoxia, since no signs of vascular damage are found in the brains of BDV-infected rats (Bilzer and Stitz 1994). Treatment of BDV-infected rats with anti-CD8 monoclonal antibodies reduces or inhibits inflammation and prevents overt loss of brain substance (Bilzer and Stitz 1994; Planz et al. 1993). In untreated rats, necrobiotic changes of brain cells are found even in the earliest stages of the disease and neuronal cell loss is a prominent feature of BD (Carbone et al. 1989; Planz et al. 1993).

Treatment of BDV-infected rats with anti-CD8 antibodies results in decreased expression of MHC class I but has no effect on MHC class II expression. Since coexpression of MHC class I antigen in association with virus-specific proteins renders cells as targets for cytotoxic CD8+ T lymphocytes (reviewed in Zinkernagel and Doherty 1979), the observation that BDV-infected rats have CD8+ infiltrates and enhanced MHC class I expression provides a possible explanation for immunopathology in this system. Recent experiments employing syngeneic and allogeneic BDV-infected target cells and lymphocyte preparations isolated from the brain of BDV-infected rats showed evidence for the activity of virus-specific classical CTLs (Planz et al. 1993). These experiments also showed weak evidence for the presence of MHC class II-restricted CTLs among lymphocytes isolated directly from the brain of BDV-infected rats. Though a cytotoxic BDV-specific CD4+ T cell line has been described (Richt and Stitz 1992), several lines of evidence suggest that the effect of MHC class II-restricted killer cells is likely to be limited; first, whereas antiviral activity of classical MHC class I-restricted virus-specific T cells has been found for various virus infections in numerous reports (Byrne and Oldstone 1984; Baenziger et al. 1986; Luckacher et al. 1985; Sethi et al 1983; Zinkernagel and Doherty 1979), in none of the virus infections in which MHC class II-restricted CTL activity was elicited in vitro (Browning et al. 1990; Hioe and Hinshaw 1989; Jacobson et al. 1984; Kaplan et al. 1984; Luckacher et al. 1984; Yasukawa and Zarling 1984), has there been direct evidence for an antiviral effector mechanism in vivo; second, MHC class II antigen has not been detected on neurons in BDV-infected rats (Bilzer and Stitz 1994); third, in contrast to studies using antibodies to CD8+ cells, in vivo treatment with monoclonal antibodies directed against CD4+ T cells did not prevent significant loss of brain tissue (Planz et al. 1993; Bilzer and Stitz 1994). However, formally virus specific CD4+ CTLs should not be discounted entirely. If operative in vivo, virus-specific CD4+ CTLs could participate in the pathogenesis by direct action on MHC class II bearing-astrocytes (Planz et al. 1993).

It appears that, in addition to the importance of CD8+ T cells in triggering a delayed type hypersensitivity reaction in the brain, CD8+ T cells are clearly important in brain tissue destruction after BDV infection and may be the critical effector cells responsible for organ atrophy chronic debility and dementia (Planz et al. 1993; Bilzer and Stitz 1993). The importance of CD8+ cells is further supported by the finding that adoptive transfer of a BDV-specific CD4+ T helper cell line does not induce encephalitis and BD in recipients in which CD8+ T cells have been functionally blocked by specific monoclonal antibodies (Planz et al. 1995).

5 Cytokine Expression in the Central Nervous System of Borna Disease Virus-Infected Rats

Although the precise role that inflammatory cells play in CNS pathology is the subject of ongoing investigation, it is well established that leukocyte populations have the ability to generate proinflammatory cytokines (QUAGLIARELLO et al. 1991), neurotoxin (GIULIAN et al. 1990) or reactive oxygen intermediates (NATHAN 1992). It is believed that proinflammatory cytokines produced by immune cells are particularly important in the process of neuronal destruction (QUAGLIARELLO et al. 1991; SELMAJ 1992). Recently, it has been convincingly shown that cytokines not only play a central role in modulating immune responses and inflammatory reactions, but can also have direct cytotoxic effects. For example, intracisternal injection of recombinant interleukin-1 (IL-1) or tumor necrosis factor (TNF) in rats induces meningitis and blood-brain barrier damage, and the two cytokines are synergistic in inducing these effects (QUAGLIARELLO et al. 1991). TGF-β2, another potent cytokine, leads to inflammation with accumulation of monocytes and to a lesser extent lymphocytes and macrophages if given locally (ALLEN et al. 1990). BD in rats represents a powerful animal model for studying the role of cytokines in neurological disorders induced by virus infection. In a recent study reverse transcriptase PCR was used to compare the expression of BDV RNA with that of several cytokine mRNAs in the brain of BDV-infected rats (SHANKAR et al. 1992) BDV RNA was first detected in the olfactory bulb of the intranasally infected rats at 6 days postinfection; at 14 days high levels of BDV RNA were found in many brain regions, and at 26 days BDV RNA was also present in the eye, nasal mucosa and facial skin. In the chronic phase of the disease, BDV RNA was identified in many peripheral organs.

Analysis of brain tissue for the presence of cytokine mRNAs revealed that the mRNA levels of IL-6, TNF α and IL-1α had increased sharply at 14 and 26 days postinfection. These cytokine mRNAs reached maximum levels at the peak of inflammatory reactions and decreased dramatically in the chronic phase of the disease. Although IL-2 mRNA was found in normal rat brain, it was markedly increased in BDV-infected brains at 14 days postinfection. Expression of IFN γ mRNA, which was not observed in normal rat brain, was detected at 14 days postinfection and reached a maximum level at 38 days. IFN-γ mRNA expression was coordinated with expression of CD4 and CD8 mRNAs consistent with the observation that both CD4+ and CD8+ T cells are induced in the early stage of BDV infection.

The finding that the levels of cytokine mRNAs correlated with the degree of inflammatory reactions and severity of neurological disease suggested that the production of certain proinflammatory cytokines may contribute to neurological disease (SHANKAR et al. 1992). The mechanisms by which cytokines are involved in the disease process are not clear. However, it has recently been proposed for experimental allergic encephalitis that neurospecific T cells recruit activated inflammatory cells through the action of cytokines such as TNF-α and -β and

IFN-γ (RUDDLE et al. 1990; SIMMONS and WILLENBORG 1990). In addition, these cytokines can prime macrophages for the production of inducible reactive oxygen intermediates and reactive nitrogen intermediates (DING et al. 1988), which are thought to play an important role in cytotoxicity (NATHAN 1992).

In a recent study, it was demonstrated that mRNA for inducible nitric oxide synthase (iNOS), which is not detectable in normal brain tissue, is up-regulated in the brain of rats infected with BDV (KOPROWSKI et al. 1993). Interestingly, the levels of iNOS mRNA correlated not only with the degree of neurological involvement and CNS inflammation, but also with the levels of TNF-α, IL-1α and IL-6 mRNA (SELMAJ 1992), potential mediators of iNOS expression (NATHAN 1992). These observations support the hypothesis that certain cytokines, such as TNF-α and IL-1α may participate in the inflammatory process by triggering infiltrating macrophages to generate NO (NATHAN 1992). To this point we have focused on damaging effects of cytokines. However, since IFN-γ mRNA levels were still highly elevated in the chronic phase of BD, it is likely that IFN-γ acts to reduce inflammation and to ameliorate neurological signs during the transition from the acute to the chronic phase of infection. In this context, it has been shown that IFN-γ has a strong synergistic effect in the TNF- and IL-1-mediated induction of manganese superoxide dismutase, a mechanism which is implicated in the protection of healthy cells from toxicity of O_2^- during an immune response (HARRIS et al. 1991). Since IL-1 enhances nerve growth factor synthesis during injury, it may be beneficial in nerve trauma (LINDHOLM et al. 1987). Furthermore, it has been demonstrated that TNF-α, INF-γ and IFN-β can exert antiviral activity in nervous tissue cells (SCHIJNS et al. 1991).

Additional in vivo studies with inhibitors of cytokines will be necessary to reveal the beneficial and detrimental attributes of cytokine expression during BD and other inflammatory diseases of the CNS.

6 Conclusion

We have tried to review a rapidly evolving field, the immunobiology of BD. While interactions between various components of the cellular immune response and between the cellular immune system and lymphokines are complex, studies in BD using methods for immunosuppression and immunomodulation with various drugs including T cell-specific monoclonal antibodies, investigations in persistently infected newborn rats and analysis of virus and virus-specific antigens and lymphokines have already resulted in major advances in understanding the basis of this immune-mediated disease. Experimental BD is an excellent model for studying the immunopathogenesis of viral infections. It also holds promise for studying such basic processes such as the induction of tolerance to foreign antigens. BD also represents a unique model for studying in vivo cytotoxicity exerted by classical CD8+ CTLs. The recent observation that BDV can be detected in macrophages indicates that BD might also serve as another virus model for phenomena observed in lentivirus infections e.g. maedi-visna in sheep and HIV in humans.

Acknowledgment. This review describes work that has been supported by grants from the Deutsche Forschungsgemeinschaft (Forschergruppe "Pathogenitätsmechanismen von Viren", Bi-323/2-2 and Sti 71/2-1) and the EC-network "Pathogenesis of subacute and chronic inflammatory diseases of the central nervous system (L.S.) and NIH NS28599 and MH48948 (K.M.C). LS is a recipient of a Herrmann- and Lilly-Schilling Professorship for Theoretical and Clinical Medicine.

References

Allen JB, Manthey CL, Hand AR, Ohura K, Ellingsworth L, Wahl SM (1990) Rapid onset synovial inflammation and hyperplasia induced by transforming growth factor a. J Exp Med 171: 231

Baenziger J, Hengartner H, Zinkernagel RM, Cole GA (1986) Induction or prevention of immunopathological disease by cloned cytotoxic T cell lines specific for lymphocytic choriomeningitis virus. Eur J Immunol 16: 387–393

Bautista JR, Schwartz GJ, Moran TH, Carbone KM (1994) Early and persistent abnormalities in size, drinking behaviors, and spontaneous locomotor activity in rats with neonatally acquired Borna disease virus infection. Brain Res Bull 34: 31–40

Bilzer T, Stitz L (1993) Brain cell lesions in Borna disease are mediated by T cells. Arch Virol Suppl 7: 153–159

Bilzer T, Stitz L (1994) Immune-mediated brain atrophy: CD8+ T cells cause tissue destruction during Borna disease. J Immunol 153: 818–823

Bilzer T, Lipkin WI, Planz O, Stitz L (to be published) Tropism of Borna disease virus. Distribution of virus-specific nucleic acid and antigen in adult and immunocompromised rats.

Browning MJ, Huang AS, Reiss CS (1990) Cytolytic T lymphocytes from the BALB/c-H-2dm2 mouse recognize the vesicular stomatitis virus glycoprotein and are restricted by class II MHC antigens. J Immunol 145: 985–994

Byrne JA, Oldstone MBA (1984) Biology of cloned cytotoxic T lymphocytes specific for lymphocytic choriomeningitis virus : clearance or virus in vivo. J Virol 51: 682–686

Caplazi P, Waldvogen A, Stitz L, Braun U, Ehrensperger F (1994) Borna disease in naturally infected cattle. J Comp Path 111: 65–72

Carbone KM, Duchala CS,Griffin JW, Kincaid AL, Narayan O (1987) Pathogenesis of Borna disease in rats : Evidence that intra-axonal spread is the major route for virus dissemination and the determinant for disease incubation. J Virol 61: 3431–3440

Carbone KM,Trapp BD,Griffin JW, Duchala CS, Narayan O (1989) Astrocytes and Schwann cells are virus-host cells in the nervous system of rats with Borna disease. J Neuropathol Exp Neurol 48: 631–644

Carbone KM, Moench TR, Lipkin WI (1991a) Borna disease virus replicates in astrocytes, Schwann cells and ependymal cells in persistently infected rats: location of viral genomic and messenger RNAs by in situ hybridization. J Neuropathol Exp Neurol 50: 205–214

Carbone KM, Park SW, Rubin SA, Waltrip RW, Vogelsang GB (1991b) Borna disease : association with a maturation defect in the cellular immune response. J. Virol 65: 6154–6164

Cervos-Navarro J, Roggendorf W, Ludwig H, Stitz L (1981) The encephalitic reaction in Borna disease virus infected rhesus monkeys (in German). Verh Dt Ges Pathol 65: 208–212

Cole GA, Nathanson N, Prendergast RA (1972) Requirement for theta-bearing cells in lymphocytic choriomeningitis virus-induced central nervous system disease. Nature 238: 335–337

Danner K, Heubeck D, Mayr A (1978) In vitro studies on Borna disease. I. The use of cell cultures for the demonstration, titration and production of Borna virus. Arch Virol 57: 63–75

Deschl U, Stitz L, Herzog S, Frese K, Rott R (1990) Determination of immune cells and expression of major histocompatibility complex class II antigen in encephalitic lesions of experimental Borna disease. Acta Neuropathol (berl) 81: 41–50

Ding AH, Nathan CF, Stuehr DJ (1988) Release of reactive nitrogen and oxygen intermediates from mouse peritoneal macrophages. Comparison of activating cytokines and evidence for independent production. J Immunol 141: 2407–2412

Dittrich W, Bode L, Ludwig H, Kao M, Schneider K (1989) Learning deficiencies in Borna disease virus-infected but clinically healthy rats. Biol Psychiatry 26: 818–828

Doherty PC, Zinkernagel RM (1974) T cell-mediated immunopathology in viral infections. Transplant Rev 19: 89–120

Doherty PC, Dunlop MBC, Parish CR, Zinkernagel RM (1976) Inflammatory process in murine lymphocytic choriomeningitis is maximal in H-2K or H-2D compatible interactions. J. Immunol 117: 187–190

Duffy P, Wolf J, Collins G (1974) Possible person-to-person transmission of Creutzfeldt-Jakob disease. N Engl J Med 299: 692–694

Fierz W, Endler B, Rese K, Wekerle H, Fontana A (1985) Astrocytes as antigen-presenting cells. I. Induction of la antigen expression on astrocytes by T cells via immune interferon and its effect on antigen presentation. J Immunol 134: 3785–3791

Fleischer B (1984) Acquisition of specific cytotoxic activity by human T4+ T lymphocytes in culture. Nature 308: 365–367

Fontana A, Fierz W, Wekerle H (1984) Astrocytes present myelin basic protein to encephalitogenic T cell-lines. Nature 307: 273–276

Fontana A, Frei K, Bodmer S, Hofer E, Schreier MH, Palladino MAJr, Zinkernagel RM (1989) Transforming growth factor α inhibits the generation of cytotoxic T cells in virus infected mice. J Immunol 143: 3230–3234

Gajdusek DC, Gibbs CJ Jr (1971) Transmission of two subacute spongiform encephalopathies of man (Kuru and Creutzfeldt-Jakob disease) to New World monkeys. Nature 230: 588–591

Giulian D, Vaca K, Noonan CA (1990) Secretion of Neurotoxins by mononuclear phagocytes infected with HIV-1. Science 250: 1593–1596

Grant I, Atkinson JH (1990) Neurogenic and psychogenic behavioral correlates of HIV infection. In: Waksman BH (ed) Immunologic mechanisms in neurologic and psychiatric disease. Raven, New York, pp 291–304

Halloran PF, Urmson J, Farkas S, Meide van der P., Autenried P (1989) Regulation of MHC expression in vivo. IFN-α/β inducers and recombinant IFN modulate MHC antigen expression in mouse tissue. J Immunol 142: 4241–4247

Harris CA, Derbin KS, Hunte-McDonough B, Krauss MR, Chen KT, Smith DM, Epstein LB (1991) Manganese superoxide dismutase is induced by IFN-γ in multiple cell types: synergistic induction of IFN-γ and tumor necrosis factor or IL-1. J Immunol 147: 149–154

Herzog S, Rott R (1980) Replication of Borna disease virus in cell culture. Med Microbiol Immunol 168: 153–158

Herzog, S, Kompter C, Frese K, Rott R. (1984) Replication of Borna disease virus in rats: age-dependent differences in tissue distribution Med. Microbiol Immunol 173: 171–177

Herzog S, Wonigeit K, Frese K, Hedrich HJ, Rott R (1985) Effect of Borna disease virus infection in athymic rats. J Gen Virol 66: 503–508

Hioe CE, Hinshaw VS (1989) Induction and activity of class II-restricted, Lyt-2 cytolytic T lymphocytes specific for the influenza H5 hemagglutinin. J Immunol 142: 2482–2488

Hirano N, Kao M,Ludwig H (1983) Presistent, tolerant or subacute infection in Borna disease virus infected rats. J Gen Virol 64: 1521–1530

Hope J, Reekie LJD, Hunter N, Multhaup G, Beyreuther K et al. (1988) Fibrils from brains of cows with new cattle disease contain scrapie-associated protein. Nature 336: 390–392

Irigoin C, Rodriguez EM, Heinrichs M, Frese K, Herzog S, Oksche A, Rott R (1990) Immunocytochemical study of the subcommisural organ of rats with induced postnatal hydrocephalus. Exp Brain Res 82: 384–392

Jacobson S,Richert JR, Biddison WE, satinsky A, Hartzman RJ, McFarland HF (1984) Measles virus-specific T4+ human cytotoxic T cell clones are restricted by class II HLA antigens. J Immunol 133: 754–757

Jahnke U, Fischer EH, Alvord EC (1985) Sequence homology between certain viral protein and proteins related to encephalomyelitis and neuritis. Science 229: 282–284

Kaplan DR, Griffith R, Braciale VL, Braciale TJ (1984) Influenza virus-specific human cytotoxic T cell clones : heterogeneity in antigen specificity and restriction by class II MHC products. Cell Immunol 88: 193–199

Koprowski H, Zeng YM, Heber-Katz E, Frazer N, Rorke L, Fu ZF, Hanlon C, Dietzschold B (1993) In vivo expression of inducible nitric oxide synthase in experimentally-induced neurological disease. Proc Natl Acad Sci USA 90: 3024–3027

Lampson LA (1987) Molecular bases of the immune response to neural antigens. Trends Neurosci 10: 211–216

Liebert UG, Linington C, ter Meulen V (1988) Induction of autoimmune reactions to myelin basic protein in measles virus encephalitis in Lewis rats. J Neuroimmunol 17: 103–118

Lindholm D, Heumann R, Meyer M, Thomas H (1987) Interleukin-1 regulates synthesis of nerve growth factor in non-neuronal cells of rat sciatic nerve. Nature 330: 658–659

Luckacher AE, Braciale VL, Braciale TJ (1984) In vivo effector function of influenza virus-specific cytotoxic T lymphocyte clones is highly specific. J Exp Med 160: 814–826

Luckacher AE,Morrison LA, Braciale VL, Malissen B, Braciale TJ (1985) Expression of specific cytolytic activity by H-2 I region-restricted, influenza virus-specific T lymphocyte clones. J Exp Med 162: 171–187

Ludwig H, Bode L, Gosztonyi G (1988) Borna disease. A persistent virus infection of the central nervous system. Prog Med Virol 35: 107–151

Manuelidis L, Manuelidis EE (1988) Recent developments in scrapie and Creutzfeld-Jakob disease. Prog Med Virol 33: 18–98

Massa PT, Dörries R, ter Meulen V (1986) Viral particles induce Ia antigen expression on astrocytes. Nature 320: 543–546

Mims CA (1960) Intracerebral injections and the growth of viruses in the mouse brain. Br J Exp Pathol 41: 52–59

Nagashima K, Wege H, Meyermann R, ter Meulen V (1978) Corona virus induced subacute demyelinating encephalomyelitis in rats: a morphological analysis. Acta Neuropathol (Berl) 44: 63–70

Narayan O, Herzog S, Frese K, Scheefers K, Rott R (1983a) Pathogenesis of Borna disease in rats: Immune-mediated viral ophthalmoencephalopathy causing blindness and behavioral abnormalities. J Infect Dis 148: 305–315

Narayan O, Herzog S, Frese K, Scheefers K, Rott R (1983b) Behavioral disease in rats caused by immmunopathological response to persistent Borna disease virus in the brain. Science 220: 1401–1403

Nathan C (1992) Nitric oxide as a secretory product of mammalian cells. FASEB J 6: 3051–3064

Padgett BL, Walter DL, zuRhein GM, Ederoode RJ, Dessel BH (1971) Cultivation of papova-like virus from human brain with progressive multifocal leukoencephalopathy. Lancet i: 1257–1260

Palladino MA, Morris RE, Starnes HF, Levinson DA (1990) The transforming growth factor-betas: a new family of immunoregulatory molecules. Ann NY Acad Sci 593: 181–185

Planz O, Bilzer T, Sobbe M, Stitz L (1993) Lysis of MHC class I-bearing cells in Borna disease virus-induced degenerative encephalopathy. J Exp Med 178: 163–174

Planz O, Bilzer T, Stitz L (to be published) Immunopathogenic role of T cell subsets in Borna disease virus-induced progressive encephalitis.

Price RW, Brew BJ (1988) The AIDS dementia complex. J Infect Dis 158: 1079–1083

Price R, Brew B, Sidtis J. Rosenblum M, Scheck A, Cleary P (1988) The brain in AIDS: central nervous system HIV-1 infection and AIDS dementia complex. Science 239: 586–592

Price RW, Brew BJ, Rosenblum M (1990) The AIDS dementia complex and HIV-1 brain infection : a pathogenic model of virus-immune interaction. In: Waksman BH (ed) Immunologic mechanisms in neurologic and psychiatric disease. Raven, New York, pp 269–290

Quagliarello VJ, Wispelwey B, Long WJ Jr, Scheid WM (1991) Recombinant human interleukin-1 induces meningitis and blood-brain barrier injury in the rat. J Clin Invest 87: 1360–1366

Richt JA, Stitz L (1992) Borna disease virus infected astrocytes function in vitro as antigen-presenting and target cells for virus-specific CD4-bearing lymphocytes. Arch Virol 124: 95–109

Richt JA, Stitz L, Wekerle H, Rott R (1989) Borna disease, a progressivo meningoencephalomyelitis as a model for CD4+ T cell-mediated immunopathology in the brain. J Exp Med 170: 1045–1050

Richt JA, Stitz L, Deschl U, Frese K, Rott R (1990) Borna disease virus-induced meningoencephalomyelitis caused by a virus-specific CD4+ T-cell mediated immune reaction. J Gen Virol 71: 2565–2573

Richt JA, Schmeel A, Frese K, Carbone KM, Narayan O, Rott R (1994) Borna disease virus-specific T cells protect against or cause immunopathological Borna disease. J Exp Med 179: 1467–1473

Ruddle NH, Bergman CM, McGrath KM, Lingenheld EG, Grunnet ML, Padula SJ, Clark RB (1990) An antibody to lymphotoxin and tumor necrosis factor prevents transfer of experimental allergic encephalomyelitis. J Exp Med 172: 1193–2000

Schijns VECJ, van der Neut R, Haagmans DR, Schellekens H, Horzinek MC (1991) Tumor necrosis factor-α, interferon-γ and interferon-α exert antiviral activity in nervous tissue cells. J Gen Virol 72: 809–815

Selmaj KW (1992) The role of cytokines in inflammatory conditions of the central nervous system. Semin Neurosci 4: 221–229

Sethi KK, Omata K, Schneweis KE (1983) Protection of mice from fatal herpes simplex type I infection by adoptive transfer of cloned virus-specific and H-2 restricted cytotoxic T lymphocytes. J Gen Virol 64: 443–447

Shankar V, Kao M, Hamir AN, Sheng H, Koprowski H, Dietzschold B (1992) Kinetics of virus spread and change in levels of several cytokine mRNAs in the brain after intranasal infection of rats with Borna disease virus. J Virol 66: 992–998

Sierra-Honigmann AM, Rubin SA, Estefanous MG, Yolken RH, Carbone KM (1993) Borna disease virus in peripheral blood mononuclear and bone marrow cells of neonatally and chronically infected rats. J Neuroimmunol 45: 31–36

Simmons RD, Willenborg DO (1990) Direct injection of cytokines into the spinal cord causes autoimmune encephalitis-like inflammation. J Neurol Sci 100: 37–42

Stephenson JR, terMeulen V (1979) Subacute sclerosing panencephalitis: Characterization of the etiological agent and its relationship to morbilliovirus. In: Tyrell DAJ (ed) Aspects of slow and persistent virus infections. Kluwer Academic, Norwell, pp 61–85

Stitz L (1992) Induction of antigen-specific tolerance by cyclosporin A. Eur J Immunol 22: 1995–2001

Stitz L, Rott R (1994) Borna disease virus. In: Webster RG, Garoff A (ed) Encyclopedia of Virology. Academic, New York, Vol 1: 149–154

Stitz L, Krey H, Ludwig H (1980) Borna disease in rhesus monkeys as a model for uveo-cerebral symptoms. J Med Virol 6: 333–340

Stitz L, Soeder D, Deschl U, Frese K, Rott R (1989) Inhibition of immune-mediated meningoencephalitis in persistently Borna disease virus infected rats by Cyclosporine A. J Immunol 143: 4250–4256

Stitz L, Schilken D, Frese K (1991a) Atypical dissemination of the highly neurotropic Borna disease virus during persistent infection in cyclosporine A-treated, immunosuppressed rats. J Virol 65: 457–460

Stitz L, Planz O, Bilzer T, Frei K, Fontana A (1991b) Transforming growth factor-β modulates T cell-mediated encephalitis caused by Borna disease virus. Pathogenic importance of CD8+ cells and suppression of antibody formation. J Immunol 147: 3581–3586

Stitz L, Sobbe M, Bilzer T (1992) Preventive effects of early anti-CD4 or anti-CD8 treatment on Borna disease in rats. J Virol 66: 3316–3323

Stitz L, Bilzer T, Richt JA, Rott R (1993) Pathogenesis of Borna disease. Arch Virol Suppl 7: 135–151

Suzumura A, Lavi E, Weiss SR, Silberberg DH (1986) Coronavirus infection induces H-2 antigen expression on oligodendrocytes and astrocytes. Science 232: 991

Tedeschi B, Barrett JN, Keane RW (1986) Astrocytes produce interferon that enhances the expression of H-2 antigens on a subpopulation of brain cells. J Cell Biol 102: 2244–2253

ter Meulen V, Stephenson JR (1983) The possible role of viral infections in MS and other related demyelinating disease. In: Hallpike JF, Adams CWM, Tourtellotte WW (eds) Multiple sclerosis. Chapmann and Hall, London, pp 241–274

Vitetta ES, Capra D (1978) The protein products of the murine 17th chromosome. Genetics and structure. Adv Immunol 26: 147–193

Wahl SM, McCartney-Francis N, Mergenhagen SE (1989) Inflammatory and immunoregulatory role of TGF-β. Immunol Today 10: 258–262

Weimer LP, Herdon RM, Narayan O, Johnson RT (1972) Further studies of simian virus-40-like virus isolated from human brain. J Virol 10: 147–152

Wilesmith JW, Wells GAH, Cranwell MP, Ryan JBM (1988) Bovine spongiform encephalopathy: epidemiological studies. Vet Rec 123: 638–644

Wolinski JS (1990) Subacute sclerosing panencephalitis, progressive panencephalitis, and multifocal leukoencephalopathy. In: Waksman BH (ed) Immunologic mechanisms in neurologic and psychiatric disease. Raven, New York, pp 259–268

Wong GHW, Bartlett PF, Clark-Lewis I, Battye F, Schrader JW (1984) Inducible expression of H-2 and Ia antigens on brain cells. Nature 310: 688–691

Yamada M, Zurbriggen A, Fujinami RS (1991) Pathogenesis of Theiler's murine encephalomyelitis virus. Adv Virus Res 39: 291–320

Yasukawa M, Zarling JM (1984) Human cytotoxic T cell clones directed against herpes simplex infected cells. I. Lysis restricted by HLA class II MB and DR antigens. J Immunol 133: 422–427

Zinkernagel RM, Doherty PC (1979) MHC-restricted cytotoxic T cells: Studies on the biological role of polymorphic major transplantation antigens determining T cell restriction-specificity, function and responsiveness. Adv Immunol 27: 52–142

Behavioral Disturbances and Pharmacology of Borna Disease

M.V. Solbrig[1], J.H. Fallon[2], and W.I. Lipkin[1, 2]

1 Introduction . . . 93
2 Behavioral Observations . . . 93
3 Pharmacology and Neurochemistry . . . 95
4 Neuroanatomy . . . 97
5 Borna Disease Virus and Human Diseases . . . 99
References . . . 101

1 Introduction

Whether Borna disease virus (BDV) or related infectious agents cause human disease remains controversial (see Bode and Amsterdam, this volume). Nonetheless, the Borna disease (BD) system has unique features that indicate it will be of utility in elucidating the pathogenesis of human neuropsychiatric diseases. Animals infected with BDV have characteristic abnormalities in social and motor behaviors. The biologic basis for these behavioral disturbances has begun to emerge through studies in rats experimentally infected with BDV. This chapter will have three objectives: (1) describe the spectrum of behavioral disturbances in natural and experimental BD in several host species; (2) define the neuroanatomy, neuropharmacology and neurochemistry of BD in the rat system (3) draw parallels between BDV-induced behavioral syndromes and human diseases.

2 Behavioral Observations

Excitability and hyperactivity, together with movement and posture disorders, are consistent clinical features in both natural and experimental disease.

[1] Department of Neurology, University of California, Irvine, CA 92717, USA
[2] Department of Anatomy and Neurobiology, University of California, Irvine, CA 92717, USA

Early in infection, horses, the most frequent natural host for BDV, have erratic excited behavior. They may ram their heads against walls or throw themselves backward. These behaviors almost certainly led to the initial description of BD in the 1800s as a "brain disease of horses causing agitation" (ZWICK 1939). Movement disorders are a prominent feature of the disease. Horses can have facial twitching, blepharospasm, abnormal movements of the tongue, lip curling, grimacing, vacuous chewing, teeth grinding, trismus, and frequent head nodding. Wide-based stance and tonic deviation of the head and neck have also been reported. Horses may also show a proclivity to press against walls or other objects, a behavior that has also been observed in BDV-infected rats (described below). Later in infection, horses become anorexic, inattentive and somnolent. The disease is typically fatal (LUDWIG and THEIN 1977). Similar disturbances in behavior are seen in sheep, the other natural host for BDV. Initial symptoms of hyperexcitability, anorexia, head pressing and separation from the flock, are followed by depression, persistent head and neck deviation, and inanition (WAELCHLI et al. 1985).

Experimental infection of prosimians and primates results in a more protracted illness with prominent disturbances in sexual and other social behaviors. Infected *Tupaia glis*, tree shrews, have been observed extensively. In the acute phase of the disease, solitary females run restlessly in their cages, avoiding usual behaviors such as jumping and climbing. When housed in groups, animals huddle together in what appears to be an abnormal seeking of physical contact. Weakness and abnormal postures "similar to crouching or torticollis" have been noted in the later phase of the disease. Many animals have abnormal ingestive behaviors, showing bulemia without weight gain. Shrews caged in mating pairs have disturbed social interactions. Females, rather than males, initiate mating behaviors. The frequency and duration of acceptance of sexual partners seems to markedly increase, but animals fail to reproduce. (SPRANKEL et al. 1978) Rhesus monkeys infected with BDV have an apparent progressive frontal lobe syndrome. Early in the course of the disease, monkeys are excited, active, disinhibited and aggressive, often attacking veterinary staff. In later disease, animals become apathetic, passive and refuse food and water (STITZ et al. 1980; LUDWIG et al. 1988).

Rodents are readily infected with BDV and have been the most commonly used model systems for studying the pathogenesis of BD. Initial reports indicated that BDV caused only persistent, asymptomatic infection in mice (KAO et al. 1984). However, recently, a mouse-adapted strain has been found to produce a hyperactive and aggressive syndrome in adult mice (RUBIN et al. 1993).

In rats, clinical features of BD vary with the host immune status. An extensive review of the role of the immune system in BD is presented elsewhere in this volume (see Stitz et al.). For purposes of this chapter, we will describe results of infection in two broad categories, rats infected as immunocompetent adults, and rats infected as neonates.

Within 2–3 weeks of intracerebral infection, immunocompetent adult Lewis rats become hyperactive, aggressive, vigilant and have exaggerated startle responses. Six weeks after infection they have an extrapyramidal syndrome and show stereotyped behaviors (the continuous repetition of certain behavioral elements). Behavioral disturbances can include head bobbing with retrocollis, dystonias and flexed seated postures, motor stereotypies (sniffing, chewing, scratching, grooming), masturbation, hoarding and ingestive disturbances (hyperphagia, coprophagia, polydipsia) and self-mutilation (chewing digits or tail). Like *Tupaia glis*, when housed in groups, rats huddle together. Animals cease normal grooming behaviors and are dishevelled. A small proportion of animals (5%–10%) become obese, achieving body weights as much as 300% of normal. By 6 months postinfection rats show premature senescence (NARAYAN et al. 1983; SOLBRIG et al. 1992a, b). Intranasal infection results in a similar course but onset of disease is delayed to 4–5 weeks after inoculation.

Rats infected as neonates have subtle behavioral disturbances. They perform poorly in running mazes (situations requiring spatial and temporal information processing) and show impaired aversive learning (DITTRICH et al. 1989). The deficits may relate to frontal and hippocampal pathology. Poor performance in behavior tests requiring strong motor output response may signify deficits in "working memory," a frontal lobe function, which permits the temporary storage of information used to guide future actions.

3 Pharmacology and Neurochemistry

Some behavioral manifestations in BD may be understood as disturbances in the function of the dopamine (DA) system. The spontaneous hyperactivity observed in BDV-infected rats is similar to that seen in ventral tegmental area (VTA)-lesioned rats or amphetamine-toxic rats. Limited or incomplete destruction of the cell bodies in the VTA results in a permanent behavioral syndrome characterized by hyperactivity and a general disorganization of behavior (LEMOAL et al. 1969, 1976). D-amphetamine is an indirect dopamine (DA) agonist, increasing synaptic DA by inducing DA release and blocking reuptake (COOPER et al. 1991). It produces a syndrome of enhanced locomotion and stereotyped behavior; higher doses or chronic administration intensifies stereotypy (FRAY et al. 1980). Both VTA lesions and d-amphetamine toxicity are experimental manipulations of the midbrain dopamine system. Disturbances in this system affect activity and attention. DA neurons originating in the VTA of the mesencephalon project to the prefrontal cortex as the mesocortical system: anteromedial caudate-putamen, nucleus accumbens, other ventral striatal areas as the mesostriatal system, and amygdala and septum as the mesolimbic system (Fig. 1).

There are several lines of evidence to indicate that CNS DA disturbances may be important in mediating aspects of behavioral pathology in the adult BDV-infected rat. BDV infection causes the gradual loss of tyrosine hydroxylase-immunoreactive (DA) neurons in VTA. BDV-infected animals have increased sensitivity to d-amphetamine and apomorphine, indirect and direct DA agonists,

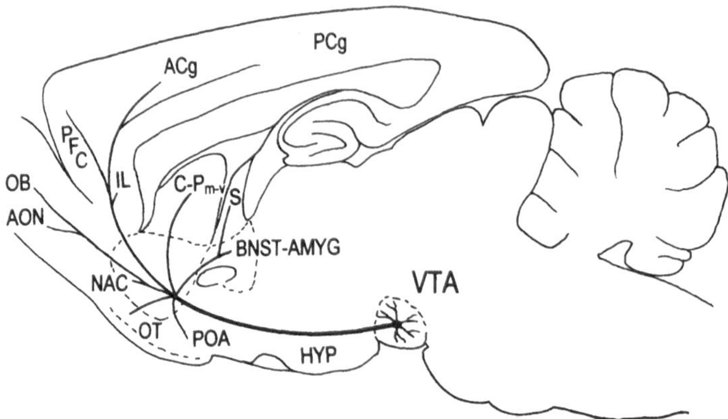

Fig. 1. Sagittal view of rat brain illustrating projections or dopamine containing neurons of the ventral tegmental area (A10). These projections include the prefrontal and limbic cortices (PFC, prefrontal orbital cortex; ACg, anterior cingulate cortex; PCg, posterior cingulate cortex; IL, infralimbic cortex), striatum ($C\text{-}P_{m\text{-}v}$, medial and ventral sectors of caudate-putamen; NAC, nucleus accumbens; OT, ventral striatal sectors of olfactory tubercle and striatal bridges), extended amygdala (BNST-AMYG, bed nucleus of the stria terminalis, amygdala), hypothalamus (POA, preoptic area; HYP, hypothalamus), septum (S), and olfactory structures (OB), olfactory bulb; AON, anterior olfactory nuclei). The projections are heaviest to striatal structures, amygdala, septum, prefrontal-orbital cortex and anterior cingulate cortex

respectively; locomotion and stereotypy are induced at lower doses and are more pronounced in BDV-infected rats than in controls. The D_1 receptor antagonist SCH23390, a drug known to stop self-injury in Lesch-Nyhan patients (Goldstein et al. 1985), aborts self-mutilation in BDV-infected rats and restores normal patterns of ambulation, grooming and consumption of food and water (Solbrig and Lipkin, unpublished observations).

There is recent neurochemical data in direct support of DA disturbances in the rat model. Regional tissue levels of DA were examined in target structures of the nigrostriatal, mesolimbic, and mesocortical dopamine systems (caudate-putamen, nucleus accumbens, olfactory tubercle and medial prefrontal cortex, respectively). The major DA metabolite, DOPAC, was also measured as an index to DA turnover and system activity. DA levels were reduced in all areas. Intriguingly, DOPAC/DA ratios in the prefrontal cortices were increased nine fold, indicating increased mesocortical DA activity and a dramatic enhancement of the prefrontal DA system. These results are consistent with partial lesions of dopamine cell groups and compensatory hyperactivity in, or unopposed positive feedback to, the remaining DA-containing cells of origin of the mesocortical system in the VTA.

DA is extensively distributed in the CNS. DA cell groups project to regions important to a wide variety of behavior and motor functions including striatal areas (caudate-putamen, nucleus accumbens, and olfactory tubercle), subcortical limbic structures (septum, olfactory tubercle, nucleus accumbens, and amygdala), neocortical and limbic cortical areas (prefrontal, cingulate, piriform, perirhinal, and entorhinal cortex), and lateral hypothalamus (FALLON and LOUGHLIN 1987). In several mammalian species, lesion, tracer and pharmacologic studies have proven a role for DA in attention and arousal; locomotor activation; dyskinesias; motor stereotypies; perseveration and behavior disorganization; ingestive, social, sexual, and reproductive behaviors; working memory; and motivational aspects of situational learning (reviewed by LEMOAL and SIMON 1991). The importance of DA to motor and social behaviors is underscored by human neuropharmacology: Parkinsonism responds to DA supplementation; schizophrenia and Tourette syndrome are treated with agents that block DA receptors; and tardive dyskinesia can be controlled with agents that deplete central dopamine and other monoamine stores.

The BD rat is a promising system for studying DA circuitry and plasticity. Further, because BDV infection causes chronic and progressive disturbances in DA systems it may more accurately represent human neuropsychiatric diseases than the fixed lesion models now in use. These features suggest that the BD system may find utility in testing psychoactive drugs. We have already found that the atypical neuroleptic, clozapine, selectively reduces activity in BD rats, but not in uninfected controls. In the long term, it may also be possible to exploit the tropism of the BDV for specific treatment of these neuropsychiatric diseases linked to DA distrubances.

4 Neuroanatomy

Though there has never been a comprehensive analysis of the distribution of BDV in the CNS over the course of disease, previous reports have highlighted specific brain regions as targets for infection. Rats infected as neonates have abnormal architecture in hippocampus and cerebellum, presumably due to defects in neuronal migration (CARBONE et al. 1991a). In adult rats, following experimental infection through intracerebral, intranasal or footpad injection, BDV proteins appear first in pyramidal neurons of hippocampus and neurons in layers 4 and 5 of frontal cortex; virus then spreads to the remaining neurons in hippocampus and the hypothalamus (CARBONE et al. 1991b). Immunocytochemical studies in horses, the natural host for BDV, have shown concentration of viral proteins in limbic cortical areas, mesencephalon, brainstem, and retina (reviewed by LUDWIG et al. 1988). Later, BDV nucleic acids are found throughout the brain in nonhomogeneous distribution: in catecholamine cell groups, basal ganglia structures, neocortex and hippocampus (SOLBRIG et al. 1992b). The time course of the pattern of infection has led to the speculation that the virus moves transsynaptically along neuronal chains (KREY et al. 1979; LUDWIG et al. 1983; CARBONE et al. 1987; MORALES et al. 1988).

We have recently begun a series of in situ hybridization experiments in order to determine the pattern of infected brain cells following intracerebral (i.c.)

injections of Borna virus into 4-weeks old rats. Seven days following intracerebral injection of 1.6 x 10^4 tissue culture infectious dose units of BDV, Giessen strain He/80, in situ hybridization signal is found in cells in restricted brain regions. These include scattered neurons in the following structures:

1. Olfactory bulb—mitral cells in the vicinity of the vomeronasal organ
2. Hippocampal area—tenia tecta and induseum grisium (hippocampal rudiment), and CA3 and CA4 regions
3. Other limbic cortices—lateral orbital and agranular insular cortex, perirhinal cortex, posterior temporal (Te2) cortex
4. Septum—lateral edge of lateral septum, medial septum
5. Diagonal band—horizontal and vertical bands
6. Amygdala—anterior amygdaloid area, basolateral and basomedial nuclei
7. Hypothalamus—supraoptic nucleus, ventral surface of hypothalamus, periventricular nucleus, adjacent medial-ventral hypothalamus
8. Thalamus—ventromedial nucleus and adjacent areas of ventral nuclei
9. Midbrain/pons—ventral tegmental area, pedunculopontine region, parabrachial nucleus, edge of ventricle IV just medial to locus coeruleus.

These preliminary studies of virus distribution at day 7 reveal some interesting functional patterns. The vomeronasal organ of the olfactory bulb is linked to detection of pheromones and to sexual behaviors. The cortical areas first infected are the ancient, limbic cortices of the hippocampal system, and prefrontal orbital/rhinal systems. The limbic system is the anatomic substrate of emotion, motivation, arousal, learning, and memory; the prefrontal system is important to cognitive attention and motor planning. Associated nuclei in the amygdala, thalamus and midbrain are also infected at this early stage. For example, the ventral tegmental area (containing the A10 dopamine neurons), basal nuclei of the amygdala and prefrontal cortex are anatomically and functionally linked together as the "prefrontal system." At later stages of disease, virus can be seen in portions of the medial caudate-putamen, nucleus accumbens, and medial thalamus. These structures are also components of the prefrontal system. Other early targets for infection include cholinergic neurons in diagonal band of Broca and the pedunculopontine region, and the supraoptic neurons of the hypothalamus. The latter are neuroendocrine neurons that secrete either oxytocin or antidiuretic hormone into the posterior pituitary. Infected periventricular and mediobasal hypothalamic neurons may mediate abnormalities in hormone function and behavior related to sexual behavior, stress, growth, circadian rhythms, homeostasis, and ingestive behaviors. Furthermore, the central gustatory circuit is infected at day 7. This taste pathway, which processes information on biologically significant chemical signals, projects from the tongue to the rostral solitary nucleus, to the parabrachial nucleus, to the ventromedial thalamic nucleus, and then to orbital/insular cortex. We are currently examining these neuroanatomical patterns of BDV infectivity at a range of time points (days 6–21 after i.c. inoculation) using in situ hybridization. These experiments may provide neuroanatomical evidence for specific patterns of neuronal susceptibility and transport of the Borna virus in the central nervous system.

5 Borna Disease Virus and Human Diseases

A number of human diseases, including Attention Deficit Disorder (ADD), schizophrenia, and Tourette's syndrome, are attended by hyperactivity, stereotypies, involuntary movements, abnormal perceptual experiences and attention deficits. Their pathogenesis is not understood and, to date, there are no satisfactory animal models for these diseases.

There are hazards to extrapolating from animal behavior to human mental states. Nonetheless, the BD rat shows extraordinary promise as a model for human motor/affective/thought disorders. Table 1 illustrates parallels between the BD rat model system and human neuropsychiatric syndromes with respect to disturbances in behavior and neurochemistry. Though the BD rat system has features in common with a number of human neuropsychiatric syndromes rather than a single psychiatric disease, the model is still persuasive. Identical symptoms are included within the diagnostic boundaries of several psychiatric diseases. For example, psychosis may be a feature of schizophrenia and bipolar-depression; hyperactivity is seen in ADD, sometimes in schizophrenia. Even designated diagnostic categories such as schizophrenia may refer to a variety of biologic conditions; schizophrenia is a heterogeneous state with various subtypes and pharmacologic responses.

Should a DA hypothesis dominate our thinking about the behavior symptoms of BD? DA is a common element in prefrontal, limbic circuits and hypothalamic targets of the chemoreceptor systems affected in BD. In addition, DA pathways can be implicated at various levels in the disease process. However, BDV does not only infect DA neurons. The extent to which behavioral abnormalities in BD reflect DA cell disturbances remains to be determined. Due to the strength of behavioral and pharmacological bioassays for DA systems, the status of DA activity may be the key to relationships among aminergic, peptidergic, and amino acid transmissions. A general statement is more useful. The earliest and most heavily infected areas in BD are systems critical or permissive for cognitive, emotional, social, and appetitive behaviors: this observation should be exploited in the design and testing of psychotherapeutic agents.

Acknowledgments. MVS wishes to thank Dr. George Koob, The Scripps Research Institute, La Jolla, CA for discussions, guidance, and equipment support for all behavioral pharmacology experiments.

Table 1. Behavioral and neuropharmacological parallels between experimental Borna disease in rats and human neuropsychiatric syndromes

Borna disease	Psychiatric diseases
Clinical Features	
Abnormal motor activity stereotypies	Stereotypies[1] Compulsions[2, 3]
Irregular, jerking movements	Mannerisms (may resemble tics)[1] Tics[3] Motor perseveration[1]
Autostimulation, Masturbation	Autostimulation[4]
Self mutilation	Self mutilation[1, 4, 5, 6, 7, 8]
Hyperactivity	Hyperactivity, agitation[1,5,9]
Increased exploration	Increased curiosity[9]
Similarities to AMP toxicity	Similarities to AMP psychosis[1]
Increased response to stimuli (auditory and tactile myoclonus)	Altered sensory experience[1, 6] Abnormal perceptual responses
Isolated areas of excellence (dexterity, biped balance)	Isolated areas of performance Excellence[4]
disturbed social behaviors: dishevelled Aggressive Constant contact with cagemates or perimeter	dishevelled[1] Poor impulse control[1, 10] Abnormal object attachments[1]
Abnormal ingestive behaviors	
Hoarding Hyperphagia Coprophagia Polydipsia or anorexia	Hoarding[1] Hyperphagia[1, 11] Coprophagia[1] Polydipsia[1] Anorexia[5, 11]
Course:- decline, then midlife improvement	Course: decline, then midage improvement[1]
Neurochemistry/pharmacology	
AMP causes increased activity and stereotypy Response to atypical neuroleptic, clozapine	AMP causes increased symptoms[1,3,12] Response to clozapine[1]
DA system supersensitivity Increased sensitivity to DA agonists	DA system supersensitivity
Response to DA antagonists	Response to DA antagonists[1]
Increased activity mesocortical DA system (Increased DOPAC/DA prefrontal cortex)	Regional imbalance by PET[1]
Abnormal 5-HT system Methysergide reduces stereotypies increased 5-HIAA/5-HT corpus striatum	Abnormal 5-HT system Response to atypical neuroleptics[1] ($5HT_2$ antagonist properties) Response to 5-HT agents[2]
Neuropathology	
Monoamine and limbic circuits (?)	Monoamine and limbic circuits[1]
Pathogenesis	
Novel negative-strand RNA virus	Primary unknown[1, 2, 3, 4, 5, 6, 7, 9, 10, 11, 12] Secondary postencephalitis variant [1, 3] (influenza)

AMP, amphetamine; DA, dopamine: DOPAC, 3,4-dihydroxyphenylacetic acid; 5-HIAA, 5-hydroxyindoleacetic acid; 5-HT, serotonin.

1, Schizophrenia; 2, obsessive-compulsive disorder; 3, Tourette syndrome; 4, autism; 5, depression; 6, psychosis; 7, mental retardation, 8, Lesch-Nyhan. 9, Attention deficit disorder: 10, aggressive

References

Carbone KM, Duchala CS, Griffin JW, Kincaid AL, Narayan O (1987) Pathogenesis of Borna disease in rats: evidence that intra-axonal spread is the major route for virus dissemination and the determinant for disease incubation. J Virol 61: 3431–3440

Carbone KM, Park SW, Rubin SA, Waltrip RW 2nd, Vogelsong GB (1991a) Borna disease: association with a maturation defect in the cellular immune response. J Virol 65: 6154–6164

Carbone KM, Moench TR, Lipkin WI (1991b) Borna disease virus in astrocytes, Schwann cells and ependymal cells in persistently infected rats: location of viral genomic and messenger RNAs by in situ hybridization. Jnl Neuropathol Exp Neurol 50: 205–214

Cooper JR, Bloom FE, Roth RH (1991) The biochemical basis of neuropharmacology, Oxford University Press, Oxford, p 298

Dittrich W, Bode L, Ludwig H, Kao M, Schneider K (1989) Learning deficiencies in Borna disease virus-infected but clinically healthy rats. Biol Psychiatr 26: 818–828

Fallon JH, Loughlin SE (1987) Monoamine innervation of the cerebral cortex and a theory of the role of monoamines in cerebral cortex and basal ganglia. In: Jones EG, Peters A (eds) Cerebral cortex, vol 6. Plenum, New York, pp 41-127

Fray PJ, Sahakian BJ, Robbins TW, Koob, GF, Iversen SD (1980) An observational method for quantifying the behavioural effects of dopamine agonists: contrasting effects of d-amphetamine and apomorphine. Psychopharmacology 69: 253–259

Goldstein M. Anderson LT, Reuben R, Dancis J (1985) Self-mutilation in Lesch-Nyhan disease is caused by dopaminergic denervation. Lancet 1: 338–339

Kao M, Ludwig H, Gosztonyi G (1984) Adaptation of Borna disease virus to the mouse. J Gen Virol 65: 1845–1849

Krey H, Ludwig H, Rott R (1979) Spread of infectious virus along the optic nerve into the retina in Borna disease virus-infected rabbits. Arch Virol 61: 283–288

LeMoal M, Simon H (1991) Mesocorticolimbic dopaminergic network: functional and regulatory roles. Physiol Rev 71: 155–234

LeMoal M, Cardo B Stinus L (1969) Influence of ventral mesencephalic lessions on various spontaneous and conditioned behaviors in the rat. Physiol Behav 4: 567–574

LeMoal M, Stinus L, Galey D (1976) Radiofrequency lesion of the ventral mesencephalic tegmentum: neurological and behavioural considerations. Exp Neurol 50: 521–535

Ludwig H, Thein P (1977) Demonstration of specific antibodies in the central nervous system of horses naturally infected with Borna disease virus. Med Microbiol Immunol 163: 215–226

Ludwig H, Bode L, Gosztonyi G (1988) Borna disease: a persistent virus infection of the central nervous system. Prog Med Virol 35: 107–151

Morales JA, Herzog S, Kompter C, Frese K, Rott R (1988) Axonal transport of Borna disease virus along olfactory pathways in spontaneously and experimentally infected rats. Med Microbiol Immunol 177: 51–68

Narayan O, Herzog S, Frese K, Scheefers, Rott R (1983) Behavioral disease in rats caused by immunopathological responses to persistent Borna virus in the brain. Science 220: 1401–1403

Rubin SA, Waltrip RW 2nd, Bautista JR, Carbone KM (1993) Borna disease virus in mice: host-specific differences in disease expression. J Virol 67: 548–552

Solbrig MV, Koob GF, Lipkin WI (1992a) Movement and behavior disorders in rats with Borna disease (BD) (Expedited poster presentation at 44th American Academy of Neurology Meeting)

Solbrig MV, Koob GF, Loughlin S, Tsai W, Lipkin WI (1992b) Borna disease virus causes dopamine disturbances in rats. Society for Neuroscience 1: 665 (abstract)

Sprankel H, Richarz K, Ludwig H, Rott R (1978) Behavior alterations in tree shrews (Tupaia glis, Diard 1820) induced by Borna disease virus. Med Microbiol Immunol 165: 1–18

Stitz L, Krey H, Ludwig H (1980) Borna disease in Rhesus monkeys as a model for uveo-cerebral symptoms. J Med Virol 6: 333–340

Waelchli RO, Ehrensperger F, Metzler A, Winder C (1985) Borna disease in sheep. Vet Rec 117: 499–500

Zwick W (1939) Borna'sche Krankheit und Encephalomyelitis der Tiere. In: Gildemeister F, Haagen E, Waldmann O (eds) Handbuch der Viruskrankheiten, Vol 2. Fischer, Jena, Germany, pp 252–354

Human Infections with Borna Disease Virus and Potential Pathogenic Implications

L. Bode

1 Introduction . . . 103

2 Diagnosis . . . 105

2.1 Detection of Antibodies . . . 105

2.2 Detection of Antigen in Blood Cells . . . 110

3 Prevalence and Geographic Distribution . . . 114

4 Clinical Aspects . . . 117

5 Epidemiological Aspects . . . 125

6 Conclusion . . . 126

References . . . 127

1 Introduction

Borna disease (BD) has been described as an infectious progressive type of nonpurulent encephalomyelitis in horses and sheep (Nicolau and Galloway 1928; Seifried and Spatz 1930; Zwick 1939; Gosztonyi and Ludwig 1984b; Ludwig et al. 1985). It is known to occur sporadically in endemic areas in Central Europe (mainly Eastern and Southern Germany) (Zwick 1939; Dürrwald 1993). In a broad variety of species, from chicken to nonhuman primates, inoculation of brain homogenates from infected animals results in clinical manifestations that range from acute fatal neurologic disease to chronic subtle neurobehavioral syndromes (Ludwig et al. 1985, 1988). Immune-mediated mechanisms (leading to inflammatory lesions) have been shown to play a major role during the development of natural (Gosztonyi and Ludwig 1984b) and experimental disease (rat) (Hirano et al. 1983; Narayan et al. 1983 a, b; Richt et al. 1989, 1990; Stitz et al. 1989). In several species, including the newborn rat (Hirano et al. 1983; Narayan et al. 1983a), the mouse (Kao et al. 1984), and the hamster (Anzil et al. 1973; Ludwig et al. 1993), persistent infection has been established that presents as learning deficiencies (Bode et al. 1989; Dittrich et al. 1989). Inapparent or subtle clinical courses have

Department of Virology, Robert Koch-Institut, Bundesgesundheitsamt, 13353 Berlin, Germany

also been observed in naturally infected horses (IHLENBURG and BREHMER 1964; LANGE et al. 1987; Ludwig et al., unpublished) and sheep (MATTHIAS 1954). The broad host range and variable clinical presentations led to consideration of the possibility that humans might be infected with BDV (AMSTERDAM et al. 1987).

The causative agent, Borna disease virus (BDV), has been regarded to represent a neurotropic agent which shares some pathogenic similarities with rabies virus (GOSZTONYI and LUDWIG 1984a; GOSZTONYI et al. 1993). In contrast to rabies virus, however, BDV has as yet not been classified and precise morphological characterization is still lacking. Recent molecular approaches (reviewed in chapter by Briese) have considerably contributed to its partial characterization as a single-stranded RNA virus with a genome between 8.5 and 10 kb (DE LA TORRE et al. 1990; LIPKIN et al. 1990; VANDE WOUDE et al. 1990). The polarity of the genome has been controversial (LIPKIN et al. 1990; RICHT et al. 1991; PYPER et al. 1993); recent data, however, clearly pointed to a negative-strand genomic RNA (BRIESE et al. 1992; LIPKIN et al. 1992). The epidemiology of BD have been assessed through the identification of BDV-specific proteins which are present in the nuclei and cytoplasm of infected cells (LUDWIG and BECHT 1977; HAAS et al. 1986; LUDWIG et al. 1988). At least two major proteins (38/40 and 24 kDa) known to be encoded by the virus genome (LIPKIN et al. 1990; VANDE WOUDE et al. 1990) are responsible for the common induction of a strong humoral immune response in infected host species. The characterization of both proteins by polyclonal and monoclonal antibodies (BAUSE-NIEDRIG et al. 1991; LUDWIG et al. 1993) has stimulated the improvement of serological detection methods and facilitated the discovery of BDV infections in humans.

Apart from early interest in a certain human neurologic disease (Economo's' encephalitis), which seemed to share intriguing histopathologic similarities with natural BD in horses (SEIFRIED and SPATZ 1930), BDV has been regarded as a nonhuman infectious agent/pathogen for almost 50 years. Observations of a nonfatal behavioral syndrome in infected tree shrews (SPRANKEL et al. 1978) and studies in infected rats expressing behavioral abnormalities (NARAYAN et al. 1983b) and learning deficiencies (DITTRICH et al. 1989) demonstrated the propensity for BDV to affect limbic structures (GOSZTONYI and LUDWIG 1984a) and suggested a possible involvement of BDV in human neuropsychiatric syndromes. Previous and more recent infections of rhesus monkeys (PETTE and KÖRNYEY 1935; STITZ et al. 1980), which led to neurologic disease and behavioral changes (CERVOS-NAVARRO et al. 1981; BODE and LUDWIG 1989; Ludwig et al., unpublished), further indicated that humans might be susceptible hosts for BDV infection, too.

Considerable progress has been made in BDV epidemiology, since the first finding of BDV-specific serum antibodies in psychiatric patients (LUDWIG et al. 1988; AMSTERDAM et al. 1985; ROTT et al. 1985) and also healthy volunteers (BODE et al. 1988). In vivo tracing of human infections has been pursued by using a variety of approaches (Table 1) in extensive seroepidemiological surveys (Table 2). The purpose of this review is to summarize our current knowledge of human BDV infections, with particular emphasis on diagnostic systems for determining their prevalence and geographic distribution and the implications of these findings for epidemiology and possibly disease.

2 Diagnosis

The diagnostic tools used to study human BDV infections have necessarily been derived from analogous techniques applied to animals. Fortunately, the majority of techniques could take advantage of the strong humoral response against BDV antigens which commonly follows natural and experimental infections irrespective of disease. In all animal species tested, antibodies uniformly recognize at least the two major BDV proteins (38/40kDa and 24kDa), previously termed as s-(soluble) antigen (LUDWIG and BECHT 1977; LUDWIG et al. 1988) and recently shown to be encoded by the virus genome (LIPKIN et al. 1990; VANDE WOUDE et al. 1990). Since antibodies against s-antigen are commonly present, especially in sera of infected species, serological methods still play the most important part for in vivo diagnosis. In animals, positive sera were mainly detected by immunofluorescence (IF) focus assays (LUDWIG et al. 1985) or related techniques (PAULI et al. 1984) making use of s-antigen in infected cells. Positive sera have also been detected by western blot or immunoprecipitation (IP) (LUDWIG et al. 1988) using separated s-antigen in gels. More recently, enzyme immunoassays (EIA) were developed to partially replace the more time-consuming IF tests (BODE et al. 1990b). Antibodies to s-antigens have also been found in cerebrospinal fluids (CSFs), especially of BD horses and rabbits, but less frequent than in serum (LUDWIG and THEIN 1977; LUDWIG et al. 1977, 1988).

Antibodies against s-antigen do not neutralize infectious virus. The existence of neutralizing antibodies has been a controversial issue (LUDWIG et al. 1988; ROTT et al. 1991). However, such antibodies have recently been demonstrated by a sensitive focus and virus spread inhibition assay in sera and CSFs from all as yet susceptible animal species, with highest titers in chicken and lowest in hamsters (LUDWIG et al. 1993).

2.1 Detection of Antibodies

The methods that have been used to study immunoreactivity to BDV are summarized in Table 1. Since the first attempt in the late 1970s, two modifications of indirect IF have been widely applied. ROTT et al. (1985) published the results of an IF test of human sera that used persistently infected Madin Darby canine kidney (MDCK) cells after acetone fixation. This test detected positive human sera by nuclear fluorescence, but needed a predilution of the samples in swine serum (1:10) and a preabsorption with swine liver powder (100 mg/ml) to eliminate nonspecific staining. We developed an IF test with a double stain technique (BODE et al. 1988, 1991, 1992a) to overcome such specificity problems. This assay was performed with young rabbit brain (YRB) cells, acetone-fixed 5 days after infection. The IF double stain test confirmed the specificity of nuclear fluorescence obtained with positive human sera by comparing the staining pattern of each positive-scored sample with that of a BDV-specific monoclonal antibody (Kfu3, LUDWIG et al. 1993) in one and the same cell using different fluorescent labels (an

Table 1. Diagnostic tools in establishing proof of human Borna disease virus infection

Method	Materials	Subjects[a]	Finding	References
Antibodies in serum				
IF	G-26 cells, persistently infected, acetone-fixed	Depression and epilepsy patient	First positive case	Ludwig and Koprowski (unpublished data) in LUDWIG et al. 1988
IF	MDCK cells, persistently infected, acetone-fixed sera preabsorbed with swine liver powder	Psychiatric in/out-patients and healthy volunteers	Patients 4–5% positive; none of volunteers	AMSTERDAM et al. 1985; ROTT et al. 1985
		Psychiatric and neurologic in-patients/ healthy volunteers	Patients 4–7% positive/volunteers 1% positive	ROTT et al. 1991
IF (double stain with monoclonal antibody)	YRB cells, 5 days postinfection, acetone-fixed	Psychiatric out-patients/patients with neurologic and immunologic disorders/healthy volunteers	Psychiatric patients 1–4% positive/disorder patients 4–14% volunteers 2%	BODE et al. 1988, 1992a
EIA (double sandwich with monoclonal antibodies and native antigen)		Patients with neurologic disorders, e.g., MS; selected IF-positive and -negative samples	Similar specificity, increased sensitivity compared with double-stain IF	BODE et al. 1990b
IP	Cell lysates, persistently infected Vero cells	Patients with > 1:320 IF titers (n=7 tested)	Reactive with 38/40 kDa protein (7/7)	BODE and RIEGEL 1988; BODE et al. 1990a; BODE et al. 1992a
IP	In vitro translated 24 kDa protein	Psychiatric in-patients (n=4)	Reactive with 24 kDa protein (3/4)	VANDE WOUDE et al. 1990
WB	Extracts, persistently infected MDCK cells	Psychiatric in-patients (n=2)	Reactive with either 38/40 or 24 kDa protein	ROTT et al. 1991
WB	Purified antigen from persistently infected RK 13 cells	Psychiatric out-patients	Main reactivity with 38/40 kDa protein, less with 24 kDa; few with both proteins	FU et al. 1993

Antibodies in CSF				
IF	MDCK cells persistently infected	Seropositive psychiatric out-patients (n=5) / in-patients (n=19)	None reactive / (10/19) reactive with increased BDV antibody index	AMSTERDAM et al. 1985; BETCHER et al. 1989b
IF	YRB cells 5 days postinfection, double stain	Seropositive MS in-patients (n=8)	None reactive	Bode et al. (unpublished data) (1990–1991)
WB		Seropositive psychiatric in-patients (n=2)	Reactive with either 38/40 or 24 kDa protein	ROTT et al. 1991
Antigen in PBMs				
FACS	Anti-38/anti-24 kDa monoclonal antibodies	Acute and chronic psychiatric in-patients; follow-up samples	Some 40–50% reactive	BODE et al. 1994 (also see this volume)
Virus isolation from CSF				
Tissue culture	Embryonic rabbit brain cells	CSF from seropositive psychiatric in-patients (n=3)	Detectable antigen-positive foci 10–12 days postinfection, transient after passages; no disease in rabbits 5 months postinfection, but antibodies; tissue culture with rabbit brain as with original CSF	ROTT et al. 1991

IF, immunofluorescence; MDCK, Madin Darby canine kidney; YRB, young rabbit brain; IP, immunoprecipitation; PBMs, peripheral blood monocytes; FACS, fluorescence-activated cell sorting; EIA, enzyme immunoassay; WB, Western blot; MS, multiple sclerosis; RK13, rabbit kidney.

[a] For detailed clinical diagnoses and number of patient cohorts, see Table 2.

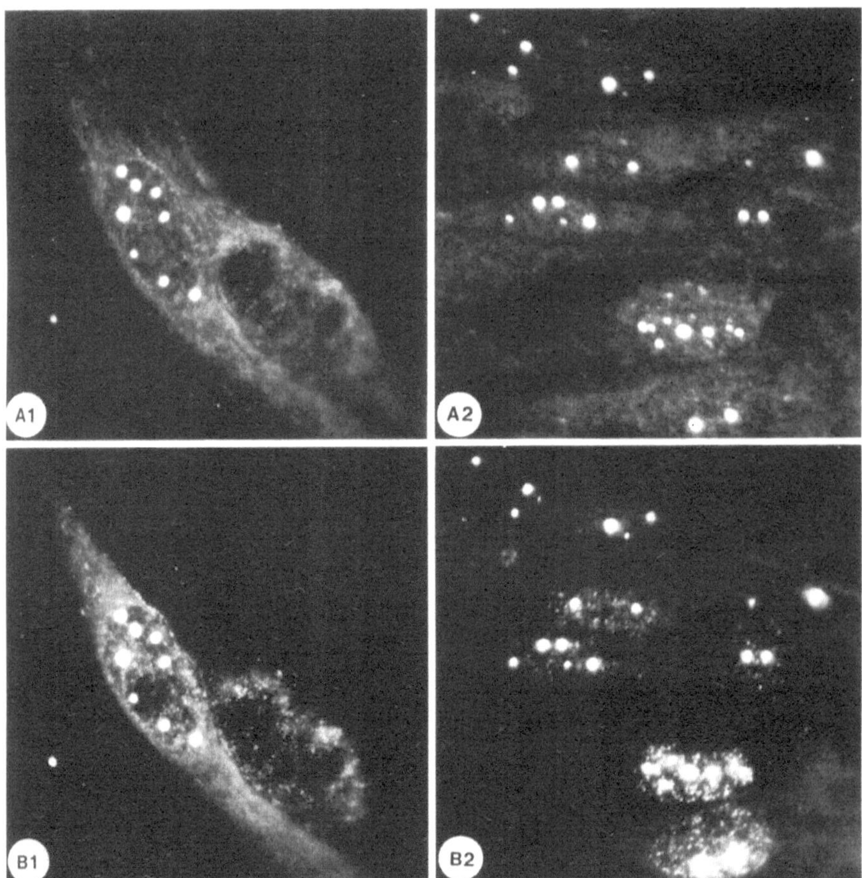

Fig. 1. Immunofluorescence (IF) of BDV-infected young rabbit brain cells with **A** human sera and **B** a BDV-specific monoclonal antibody (mab Kfu 3; LUDWIG et al. 1993). Double-stain with **A** FITC anti-human IgG/IgM; **B** TRITC anti-mouse IgG. **A1** Serum from a psychiatric patient; **A2** serum from a HIV infected patient. From BODE et al. 1992a, with kind permission of Wiley Liss. Inc., New York

example is given in Fig. 1). Any preabsorption procedures became redundant by this protocol.

It should be emphasized that the results of both types of IF assays are comparable in so far as both detected nuclear fluorescence and both found low antibody titers (1:10 to 1:80, but occasionally 1:640 or higher), irrespective of the patients' diseases (AMSTERDAM et al. 1985; ROTT et al. 1985; BODE et al. 1988; BECHTER et al. 1992c); in positive healthy volunteers, solely low titers were found (BODE et al. 1988, 1992a). Evidence has emerged to indicate that nuclear fluorescence and low IF titers are common features of natural BDV infections. Horses with acute BD have low antibody levels (LUDWIG and THEIN 1977; LANGE et al. 1987; DÜRRWALD 1993) in the same range as found in human subjects. Similarly low IF titers have been measured in sera of inapparently infected horses in Germany

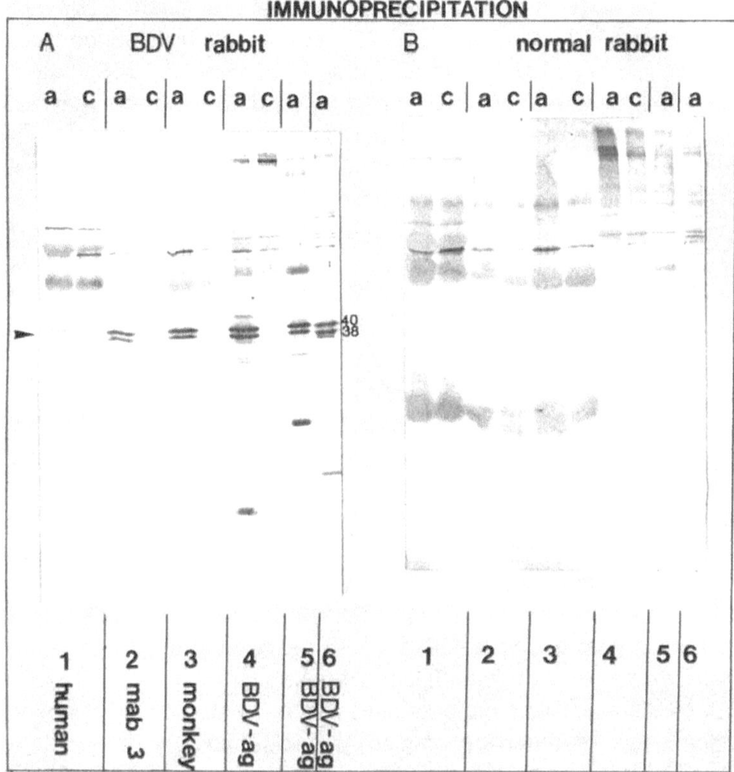

Fig. 2A, B. Immunoprecipitation of lysates from BDV-infected (*a*) and uninfected (*c*) Vero cells with: *1*, a human serum (A1, Fig. 1); *2*, monoclonal antibody Kfu 3; *3*, serum from a rhesus monkey after intracerebral infection with BDV. BDV antigen controls from different sources; *4*, Vero cells (*a, c*); *5*, rat brain, and human oligodendroglia cells. Immunostain of blotted precipitates with (**A**) BDV rabbit serum and (**B**) normal rabbit serum. From BODE et al. 1992a, with kind permission of Wiley Liss. Inc., New York

(LANGE et al. 1987) and North America (KAO et al. 1993); the latter results, however, were obtained by a less sensitive IF test (using formaldehyde-fixed baby hamster kidney cells). Interestingly, experimentally infected rhesus monkeys developed low titers (and nuclear fluorescence) after intranasal (i.n.) BDV infection (BODE and LUDWIG 1989), but higher titers and both nuclear and cytoplasmic fluorescence after intracerebral (i.c.) infection (LUDWIG et al. 1988). This lends indirect support to the generally accepted suggestion that i.n. transmission is the main route for infection in natural BD.

Methods other than IF have been employed to define BDV antigens reactive with human sera. We have demonstrated immunoreactivity with the 38/40kDa protein by an IP technique using BDV infected cell-lysates, as shown in Fig. 2 (BODE and RIEGEL 1988; BODE et al. 1990a, 1992a). IP has the advantage of catching the whole reactivity spectrum of the serum sample in question by offering almost native antigen. However, this method needs a considerably higher antibody concentration than IF, since only human sera with IF titers of at least 1:320 were found to be positive by IP (Bode et al. 1992a). There has been evidence from previous results with CSFs and sera from BD horses (Ludwig et al. 1988) that, in natural infections, humoral response against the 38/40kDa protein may be predominant. In this context, it may be important that we could not detect antibodies against the 24kDa protein in (a limited number of) human samples. However, such antibodies have been demonstrated by others in a few sera which precipitated in vitro translates of a BD cDNA clone (Vande Woude et al. 1990). Rott et al. (1991) reported that two patients' sera either reacted with the 38/40kDa or the 24kDa protein in a western blot done with extracts of persistently infected MDCK cells. Fu et al. (1993) found that a low number of patients with affective disorders reacted with both proteins by western blot, while a higher proportion reacted with either the 24kDa or the 38/40kDa protein. With the latter protein, the highest reactivity was obtained. The differences in results with these studies may be explained by differences in sensitivity of the assays employed. There is no question that a western blot performed with affinity chromatography-purified antigens (Fu et al. 1993) should detect smaller concentrations of antibodies than similar tests (western blot of IP) performed with unpurified cell extracts; in other words, it might be more likely that the reported differences in studies of human sera are due to a quantitative rather than qualitative effects. It should be noted that human antibodies seem to be solely directed against s-antigen and fail to neutralize virus (Bode and Ludwig, unpublished results).

CSFs of selected seropositive patients have also been studied for pursuing antibodies to BDV. Since intrathecally produced antibodies had been demonstrated in the CSFs of both experimentally infected rabbits (Ludwig et al. 1977, 1988) and (some but not all) horses with natural BD (Ludwig and Thein 1977; Dürrwald 1993), the presence of such antibodies could also be expected in some cases of human infection. Although evidence for BDV-specific CSF antibodies was reported in some seropositive hospitalized acute psychiatric patients (Bechter et al. 1989b), negative findings in (seropositive) psychiatric out-patients (Amsterdam et al. 1985) or multiple sclerosis (MS) in-patients (Bode et al., unpublished results) were also frequent, indicating that CSF antibodies to BDV were infrequent.

Interestingly, the attempt to isolate virus from (antibody containing) CSFs of two patients has led to transient antigen-positive foci in tissue culture and antibody production but no disease in i.c. inoculated rabbits (Rott et al. 1991). While failure to isolate virus is disappointing, it is important to consider that virus was isolated from the CSFs of only every third rabbit with BD (Ludwig et al. 1977).

2.2 Detection of Antigen in Blood Cells

In addition to pursuing antibody tests, we have tried to find new markers to use in studying human infection. Recently, we succeeded in detecting BDV-specific antigen in peripheral blood monocytes (PBMs) from horses with natural BD by specific FACS (fluorescence-activated cell sorting) analysis (Steinbach et al. 1993).

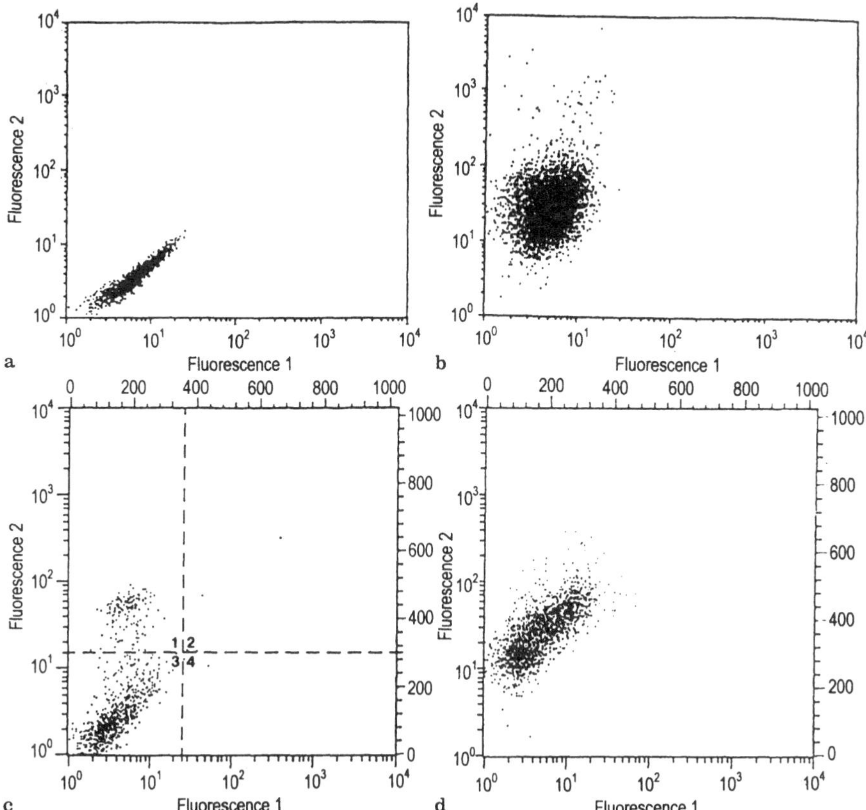

Fig. 3a–d. Flow cytometric detection of BDV-specific antigens in monocytes from naturally infected animals stained by two sets of monoclonal antibodies (mabs) (W1/W10 against 38/40 kDa, Kfu 1/2 against 24 kDa proteins; LUDWIG et al. 1993) and phycoerythrin (orange *fluorescence 2*) labeled anti-mouse IgG (Sigma). Dot-plots of monocytes from **d** a horse with acute BD, **b** a clinically healthy foal which had acquired BDV infection from its mother, after labeling with BDV-specific mabs and **a** the same cells, after labeling with control mabs; in addition, the dot-plot from monocytes of **c** a 1-week-old rat naturally infected by its mother, showing a subpopulation of positive cells after labeling with BDV-specific mabs; in **c** negative and positive cells are divided by quadrant statistics (positive cells in quadrant 1). *Fluorescence 1*, green fluorescence axis, here representing the autofluorescence of the cells. The dot-plots obtained with a Becton-Dickinson computer-assisted FACScan were kindly provided by F. Steinbach; this figure will be also published in STEINBACH et al. 1993

Flow cytometry has recently become the method of choice for cell typing. We developed a FACS assay on blood monocytes (STEINBACH and THIELE 1994), basically according to the method of CARTER (1990). Briefly, peripheral blood mononuclear cells (PBMCs) were isolated following standard procedures (BÖYUM 1968), then paraformaldehyde-fixed and treated with Triton X-100 (PAULI et al. 1984). Internal BDV antigens were stained with a mixture of two sets of monoclonal antibodies (mabs) (W1/W10 against 38/40kDa, Kfu 1/2 against 24kDa protein; LUDWIG et al. 1993) and phycoerythrin (PE)-labeled donkey anti-mouse IgG. Monocytes were identified by a computer-assisted FACScan (Becton-

Dickinson) which differentiates the cell types present in PBMC preparations according to size, extent of granularity, and fluorescence; the desired cells (PBMs) were analyzed in a dot-plot, as shown in Fig. 3. We could clearly and for the first time demonstrate that monocytes from a horse with acute BD (DÜRRWALD 1993) were stained by the specific mabs (Fig. 3d) but not by control mabs (Fig. 3a). Furthermore, the same specific staining was not restricted to monocytes from diseased animals, but could also be achieved with those of a persistently infected, but apparently healthy foal which had acquired BDV infection from its mother (Fig. 3b).

Based on these data, we now present the first flow cytometric detection of BDV antigens in monocytes isolated from patients with acute or chronic psychiatric diseases (Fig. 4). The PBMC samples were collected during two current follow-up studies and investigated in parallel with antibody tests (BODE et al. 1993a, b; Bode et al., unpublished results). This study design revealed that 40%–50% of subjects had BDV antigen in monocytes (Table 1) and 20% antibodies to BDV (Table 2). Prevalence data will be discussed later. Figure 4 depicts typical dot-plots of BDV antigen-positive monocytes. It is evident that in all three patients a subpopulation of monocytes (15%–17%), not lymphocytes (data not shown), clearly reacted positive with BDV-specific mabs (Fig. 4d–f) and not with control mabs (Fig. 4c). Proof that we dealt with monocytes but not other cells within the PBMC populations is given in Fig. 4a, b. The scattergram analysis (Fig. 4a) shows that the monocytes were gated according to their size and granularity; the nongated cells were lymphocytes. Figure 4b confirms that the gating was correct, as the majority (at least 90%) of the cells in the gate were indeed $CD14^+$ monocytes; the remaining $CD14^-$ cells were neither $CD67^+$ neutrophilic granulocytes (Fig. 4b) nor $CD14^+/CD8^+/CD19^+$ lymphocytes. Further controls

Fig. 4a–f. Flow cytometric identification of human monocytes (PBMs) within peripheral blood mononuclear cells (PBMCs) from whole blood **a, b** and detection of BDV antigens in a subpopulation of monocytes from acute **e** and chronic **d, f** psychiatric patients. **a** Scattergram of PBMCs shows the gated monocytes as compared with nongated cells below (lymphocytes), corresponding to their size (*forward scatter*) and granularity (*side scatter*). **b** Dot-plot of monocytes after double-labeling with phycoerythrin (PE) (orange, *fluorescence 2*) anti-human CD14 (IOM2) and FITC (green, *fluorescence 1*) anti-human CD 67 (ION2) monoclonal antibodies (mabs) (Immunotech) identifying the majority (~90%) of the cells as $CD14^+$ monocytes (*quadrant 1*) and excluding that the remaining $CD14^-$ cells (*quadrant 3*) are neutrophilic granulocytes (<1%, *quadrant 4*); further controls have shown that $CD14^-$ are not lymphocytes **d–f** Dot-plots of patients' monocytes after labeling with the same set of BDV-specific mabs as for monocytes from BDV-infected animals, described in Fig. 3, **c** with the same control mabs and all with PE (orange, *fluorescence 2*) anti-mouse IgG (Sigma); antigen-positive subpopulations (15%–17%) are present in **e** a seronegative acute patient, 3 weeks after onset of an episode of major depression (unipolar), **d** a seropositive chronic patient with a 6-year-history of an organic psychosyndrome leading to dementia, **f** a seronegative chronic patient with a 19-year-history of an organic mood disorder and epileptic seizures due to cerebral hemorrhage, during a depressive episode; in **d** and **f** negative and positive cells are divided by quadrant statistics; **c** shows the autofluorescence (green, *fluorescence 1* axis) of the same monocytes after labeling with control mabs and PE anti-mouse IgG. Clinical diagnoses were given (**e**) by Prof. Dr. R. Ferszt, Klinikum Steglitz, Free University and (**d, f**) by Dr. G. Arkenberg and Dr. E.H. Kang-Welberts, District Hospital Zehlendorf, Berlin. (The scattergram and dot-plots obtained with a Becton-Dickinson computer assisted FACScan were kindly provided by F. Steinbach; this figure is partially published in BODE et al. 1994) revealed that lymphocytes did not bear BDV antigen. In our patients, the proportion of positive monocytes was less than that found in BDV infected horses (Fig. 3b, d), but quite similar to that from a 1-week-old rat, naturally infected by its i.c.-infected mother (Fig. 3c). These exciting results offer the promise of a sensitive in vivo marker for BDV infection which is present in diseased and healthy animals and in psychiatric patients (BODE et al. 1994).

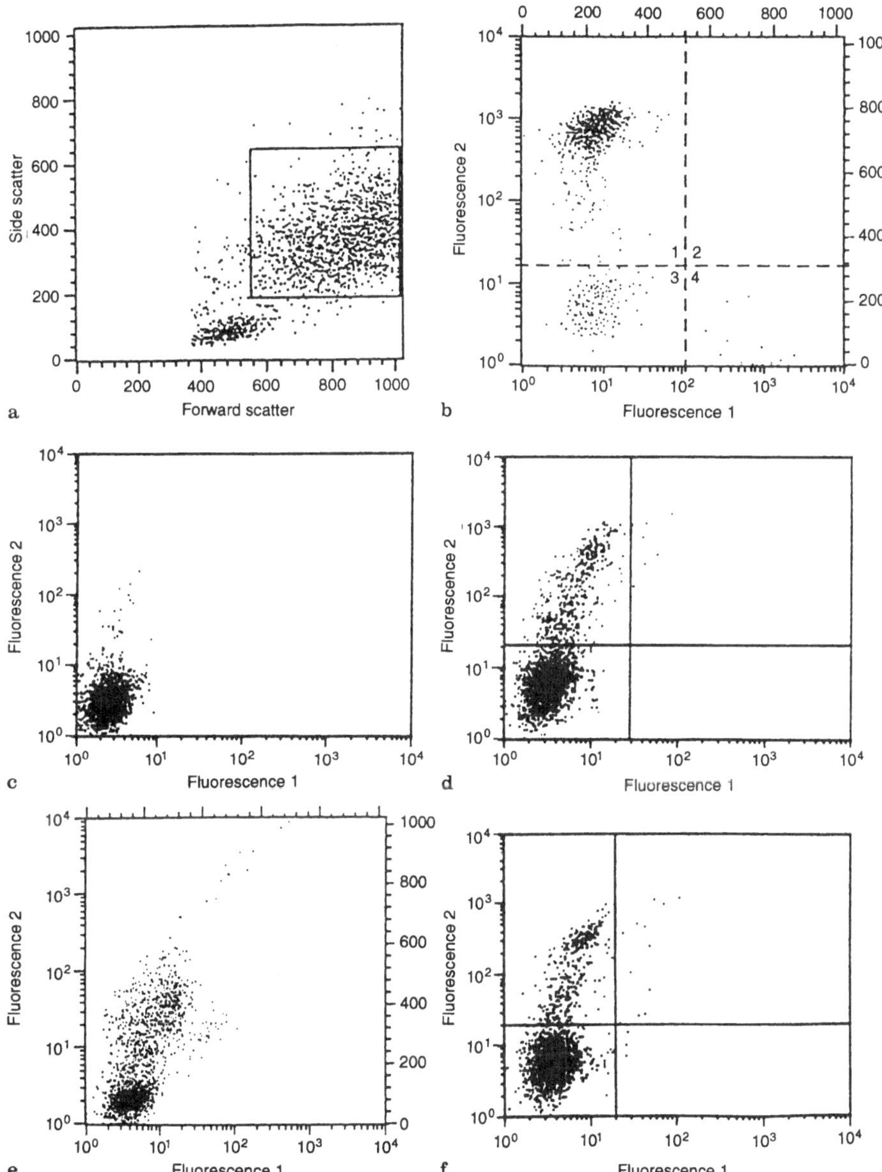

The finding of BDV antigens in monocytes should not be surprising even given that previous attempts to find such antigens have not been successful (NARAYAN et al. 1983a). Another recent study has demonstrated viral dissemination in peripheral organs, in rats immunosuppressed by cyclosporine A treatment (STITZ et al. 1991). The ability of BDV to infect cell types other than neurons (astrocytes, Schwann cells, ependymal cells) has clearly been demonstrated, especially during the chronic phase of BD in rats (LUDWIG et al. 1988; CARBONE et al. 1991; GOSZTONYI et al. 1991). Moreover, astrocytes have been shown to harbor BDV-specific antigen also throughout the course of natural BD in horses (GOSZTONYI and LUDWIG 1984b) and are susceptible to in vitro infection (RICHT and STITZ 1992). It therefore seems not unlikely that the corresponding peripheral cell type, the blood monocyte which we presented here as harboring BDV antigen, may not only function as a carrier of virus/viral antigens, but might also allow virus replication as suggested (STEINBACH et al. 1993). The failure of the previous attempt to detect antigen in rat monocytes (NARAYAN et al. 1983a) was likely due to the lower sensitivity of assays available 10 years ago. In terms of sensitivity, it seems at first sight quite amazing that molecular techniques like reverse transcriptase PCR have recently failed to detect BDV RNA in the blood of infected rats, while giving positive signals with some other nonneuronal tissues (SHANKAR et al. 1992). However, analysis of whole blood would be anticipated to be less sensitive than antigen detection in selected cells by FACS. Recently, our findings were supported by the detection of BDV RNA in PBMCs of persistently infected rats by a reverse transcriptase PCR EIA (SIERRA-HONIGMANN et al. 1993). The discovery of antigen-carrying PBMs represents an important advance in the epidemiology of BDV in human populations (BODE et al. 1993c).

3 Prevalence and Geographic Distribution

All studies on the prevalence of human infections reported so far have been based on screening of BDV serum antibodies by IF; a summary is given in Table 2, according to the clinical diagnoses. The majority of reports dealt with psychiatric and neurologic diseases because of the nature of syndromes seen in animals (see below and Tables 3, 4). The purpose of these studies was to evaluate the mean prevalences of immunoreactivity to BDV in a large spectrum of different diseases. From some thousand patients, similar data were obtained by different investigators.

Among psychiatric out-patients, a prevalence of only 2%–4.5% of antibody carriers was detected (AMSTERDAM et al. 1985; ROTT et al. 1985; AMSTERDAM et al. 1988; BODE and RIEGEL 1988; BODE et al. 1988, 1992a), whereas psychiatric in-patients expressed somewhat higher carrier rates of 4%–7% (BECHTER et al. 1987; ROTT et al. 1991; BECHTER et al. 1992c); the highest rates (10%) were found if sampling happened during onset of acute disease (BODE et al. 1991). In contrast to those screening tests (one patient/one sample), much higher seroprevalence rates of approximately 20% were found by follow-up testing (one patient/several

Table 2. Prevalence of human Borna disease virus infection

	Geographic area	Number of subjects	Number of positives[b]	(%)	Reference
Psychiatric diseases[a]					
Affective disorders[a]	USA (Philadelphia area)	265	12	(4.5)	AMSTERDAM et al. 1985 AMSTERDAM et al. 1987 ROTT et al. 1985, 1991
		up to 596	12	(2.0)	AMSTERDAM et al. 1988 BODE and RIEGEL 1988
Schizophrenia[a]		46	1	(2.2)	BODE et al. 1988 and 1992a
Chronic fatigue sydndrome (CFS)	USA (Boston area)	50	0	(0)	ROTT et al.1985
Various psychiatric diseases	Southern Germany (endemic area for horse Borna, disease)	694	4	(0.6)	
		up to 1003	68	(6.8)	BECHTER et al. 1987
		up to 2377	140	(5.9)	BECHTER et al. 1992c
Psychiatric and neurological diseases	Southern Germany, USA, Japan	sum of 5000	200–350	(4–7)	ROTT et al. 1991
Acute psychiatric patients	Eastern Germany (Berlin area)	48	5	(10.4)	BODE et al. 1991
Follow-up testing		43	10	(23.3)	BODE et al. 1992b
		up to 71	15	(21.1)	BODE et al. 1993a and b
Chronic psychiatric patients					
Follow-up testing		68	12	(17.6)	Bode unpublished, this volume
Neurologic diseases					
Multiple sclerosis[a]	Western Germany (Cologne area)	20	0	(0)	BODE and RIEGEL 1988
Multiple sclerosis (chronic progression)	Southern Germany (Munich area)	114	15	(13.2)	BODE et al. 1992a
		up to 189	25	(13.2)	Bode et al. unpublished data (1991–1992)
Nonviral encephalitis	Western Germany (Cologne area)	54	0	(0)	BODE et al. 1992a
Peripheral neuropathies		92	4	(4.3)	
Various neurologic diseases	Western Germany (Bochum area)	508	33	(6.5)	Bode et al. unpublished data (1989–1993)
	Southern Germany (endemic area)	1791	87	(4.9)	BECHTER et al. 1992c
Immunosuppressive infectious diseases					
HIV infection (asymptomatic)[a]	Eastern Germany (Berlin area)	460	36	(7.8)	BODE and RIEGEL 1988 BODE et al. 1988
		up to 1024	73	(7.1)	BODE et al. 1992a
HIV infection (lymphadenopathy)		244	34	(13.9)	BODE et al. 1992a BODE and RIEGEL 1988

Table 2. *(Continued)*

	Geographic area	Number of subjects	Number of positives[b] (%)		Reference
Epstein-Barrvirus infection		24	0	(0)	Bode et al. 1988
		up to			
(Adults)		89	1	(1.1)	Bode et al. 1992a
(Children)		160	9	(5.6)	
Multiple parasitic infections[a]	East Africa (rural area)				
(Adults)		145	10	(6.9)	Bode et al. 1992a
(Children)		48	9	(18.8)	
Controls					
Surgery patients (non-neurologic diseases)	Southern Germany (endemic area)	133	4	(3.0)	Bechter et al. 1987
		up to 569	20	(3.5)	Bechter et al. 1992c
Intravenous drug abusers[a] (HIV-seronegative)	Eastern Germany (Berlin area)	106	4	(3.8)	Bode et al. 1988
Healthy volunteers[a]	USA (Philadelphia area)	105	0	(0)	Amsterdam et al. 1985 Amsterdam et al. 1987
		up to 365	8	(2.2)	Rott et al. 1985 Amsterdam et al. 1988 Bode and Riegel 1988 Bode et al. 1988 and 1992a
	Southern Germany (endemic area)	95	0	(0)	Rott et al. 1985
		up to 1000	10	(1.0)	Rott et al. 1991
Blood donors[a]	Eastern Germany (Berlin area)	175	3	(1.7)	Bode et al. 1988 and 1992a

[a] Only out-patients as well as healthy donors indicated, all other patients were (hospitalized) in-patients.
[b] Screening of serum antibodies to Borna disease virus by immunofluorescence.

serial samples) of acute and chronic psychiatric in-patients (Bode et al. 1992b, 1993a, b; Bode et al., unpublished results).

Among in-patients with various neurologic diseases, the same frequencies of antibody carriers (4%–7%) were found as among psychiatric in-patients (Bechter et al. 1992c; Bode et al. 1992a; Bode et al., unpublished data). Interestingly, not a single BDV antibody carrier could be found among patients with nonviral encephalitis. In-patients with chronic progressive MS had a 13% prevalence of antibodies to BDV (Bode et al. 1992a); however, no seropositive MS out-patients were found (Bode and Riegel 1988). Some 14% of HIV-infected individuals with lymphadenopathy were immunoreactive (Bode et al. 1992a), whereas only 7%–8% of asymptomatic HIV carriers were BDV seropositive (Bode and Riegel 1988; Bode et al. 1988, 1992a). Up to 18% of children chronically infected with multiple blood parasites had antibodies to BDV, lending strong support to the hypothesis that chronic immunologic and/or neurologic disorders may provoke repeated reactivation events of latent BDV infection (BODE et al. 1992a).

In contrast, immunoreactivity to BDV was infrequent in healthy volunteers. Prevalence rates of approximately 2% have been reported by several investigators (AMSTERDAM et al. 1988; ROTT et al. 1991; BODE et al. 1988, 1992a). Comparably low seropositive rates of 3%–4% were also found among nonneurologic surgery patients (BECHTER et al. 1987, 1992c) and HIV-negative drug abusers (BODE et al. 1988). These studies also suggested that BDV may be a more common agent than previously thought, since seropositive individuals were detected not only in Central Europe, but also in North America, East Africa, and Japan (see Table 2).

4 Clinical Aspects

In contrast to studies in humans the pathogenic potential of BDV in animals is well documented by numerous reports (Table 3). A striking finding is the great variability in signs and clinical courses with different host species and virus strains (LUDWIG et al. 1985, 1988). As summarized in detail in Table 3, neurologic symptoms often occurred in common with behavioral abnormalities during experimental as well as natural BDV infection of almost all species. Besides disease, in every species investigated so far (except rabbit), a certain proportion of infected animals also exhibited an asymptomatic course, developing slight or no clinical signs and suggesting that inherited factors of individual hosts within

Table 3. Spectrum of neurologic disorders and behavioral changes in animals with experimental or natural Borna disease virus infection

Species	Neurologic symptoms	Behavioral changes	References
Experimental infection[a]			
Mouse	None	Not observed	KAO et al. 1984
Hamster	None	Not observed	ANZIL et al. 1973 LUDWIG et al. 1993
Tree shrew (Tupaia glis)	Nonfatal neurologic disease (only some cases):	Phase of hyperactivity	SPRANKEL et al. 1978
Solitary	hind leg ataxia, inability to jump (transient)	(exaggerated running and eating) followed by an apathic phase	
Paired	None (one exception)	No hyperactivity, but altered social behavior (exaggerated contact, docility, reversal role of sex partners)	
Rat	None		HIRANO et al. 1983
Neonatal infection		Learning deficiencies (spatial learning and memory impaired) Emotional changes (less anxiety, reduced resting)	NARAYAN et al. 1983a DITTRICH et al. 1989 BODE et al. 1989

Table 3. *(Continued)*

Species	Neurologic symptoms	Behavioral changes	References
Adult infection	Subacute/chronic meningo-encephalomyelitis/ blindness (rarely fatal; depends on virus and rat strain)	Phase of hyperactivity (aggressiveness, exaggerated eating) followed by a passive apathic phase	NARAYAN et al. 1983b; HIRANO et al. 1983
	Nonfatal obesity syndrome (some cases)	Abnormal increase of appetite	KAO et al. 1983
Rabbit	Acute encephalomyelitis/blind-ness (always fatal) ataxia, paresis of fore and hind limbs, multifocal retinopathy anorexia	Somnolence, apathy	ZWICK 1939 LUDWIG et al. 1985
Rhesus monkey			
Intracerebral route	Acute encephalomyelitis (fatal) hind leg ataxia, progressive paralysis, anorexia, retinopathy	Somnolence apathy loss of appetite	STITZ et al. 1980
Intranasal route	Acute encephalomyelitis (fatal); delayed onset of symptoms compared to the intracerebral route	excitation aggressiveness, loss of appetite	CERVOS-NAVARRO et al. 1981 BODE and LUDWIG 1989
Intraperitoneal route	None	Abnormal increase of appetite	Ludwig et al. (unpublished data)
Intranerval (ischiadicus) route[b]	Slight or no clinical signs		PETTE and KÖRNYEY 1935
Natural infection[b]			
Horse Central Europe	Acute encephalomyelitis (mainly fatal): intermittent fever, sensorial disturbances, ataxia, compulsive movements, anorexia, sometimes blindness	Somnolence, phases of excitation followed by apathy, loss of appetite	SEIFRIED and SPATZ 1930 ZWICK 1939 LUDWIG et al. 1985 DÜRRWALD 1993
Central Europe	Slight or no clinical signs		IHLENBURG and BREHMER 1964 LANGE et al 1987.
USA			KAO et al. 1993 Ludwig et al. (unpublished data)
Sheep Central Europe	Acute encephalomyelitis (mainly fatal), symptoms similar to horse disease	Somnolence, apathy	NICOLAU and GALLOWAY 1928 LUDWIG et al. 1985 DÜRRWALD 1993
	Slight or no clinical signs		MATTHIAS 1954
Cat Sweden	Acute meningoencephalo-myelitis "staggering disease" (mainly fatal), hind leg ataxia and paresis, fever, inability to retract claws, hypersensitivity to sound and light	Emotional changes, depression, increased affection, reduced appetite	LUNDGREN and LUDWIG 1993
Germany	No clinical signs associated so far		LUNDGREN et al. 1993

[a] Route of infection: intracerebrally if not indicated otherwise. Clinical course is also influenced by BDV strain, LUDWIG et al. 1993.

[b] Suggested route/transmission: intranasally.

the same species must influence the outcome of infection. The variability observed in animals may have implications for BD in humans (Bechter and Herzog 1990). The biphasic course of behavioral abnormalities shown in infected tree shrews (Sprankel et al. 1978) and rats (Narayan et al. 1983b), which was characterized by an initial hyperactive frenzied phase followed by a passive apathic phase, has been considered as highly suggestive of affective disorders in humans (Amsterdam et al. 1987). Further support for this model came from three observations: (1) that limbic structures are involved in the pathogenesis of BD both in animals and in human affective illnesses; (2) BDV was shown to exhibit a high affinity to excitatory synaptic fields in the hippocampus of rats (Gosztonyi and Ludwig 1984a); and (3) BDV induces neurotransmitter abnormalities which fit with the observed phasic course of BD in rats (Lipkin et al. 1988).

Thus, since the beginning of research on human infections, psychiatric diseases were considered to potentially be associated with BDV; a summary of reports suggesting such a relationship is given in Table 4. The majority of studies were restricted to serum antibodies as the only measurable parameter indicating evidence for possible BDV involvement. These studies found significant differences in immunoreactivity to BDV in acute, but not chronic, psychiatric diseases. For example, Bechter et al. (1987, 1989a) found significantly different prevalences of focal cerebral lesions in the white matter (examined by magnetic resonance imaging) only among seropositive compared to seronegative acute patients. As such lesions pointed to an encephalitic process, and other causes like vascular diseases, MS, and borreliosis had been ruled out, an association to BDV infection was suggested (Bechter et al. 1989a). In animals, such inflammatory lesions occur in the gray matter of the brain (Ludwig et al. 1988), which would argue against this, but such cortical lesions may not be detectable by magnetic resonance imaging (MRI). The additional detection of an increased BDV-specific antibody index in CSFs of half of the seropositive patients (10/19) (Bechter et al. 1989b) might support the hypothesis of the authors that BDV can initiate rather benign courses of encephalitis in such patients, leading to psychiatric syndromes. However, such a contributory or even initiating role of BDV, especially in affective disorders, needs further support by virus-specific parameters other than antibodies. Furthermore, it should be considered that such parameters may significantly change during acute disease and therefore only follow-up sampling may truely reflect a possible pathogenic role of the virus.

We have recently investigated a comparable number (n=70) of different cohorts of acute and chronic psychiatric in-patients by determining the presence of BDV-specific antibodies (IF method) in serum (Bode et al. 1992b, 1993a, b) and BDV antigen in PBMs (FACS method) (Bode et al. 1993c). In the acute patient cohort, the observation period was 6 weeks/patient (starting at onset of acute episode); in the chronic patients, who had a mean duration of illness of 20 ± 14 years, the observation period was 1 year; further descriptive features of the patients will be published elsewhere.

These studies supported the suggestion that BDV infection plays a role in the course of some psychiatric diseases, since both cohorts presented with

Table 4. Mental and neurologic disorders associated with human Borna disease virus infection

	BDV-specific test parameter	Suggested association	References
Psychiatric dieseases			
Nonacute episode of major depression (uni/bipolar)	Serum antibodies	Antibody carriers only among patients (4.5%), not controls	AMSTERDAM et al. 1985
Acute and chronic Major depression, (unipolar), schizophrenia, personality disorder	Serum antibodies	Focal cerebral lesions in the white matter (MRI): significant differences (seropositive/negative) only between acute (60%–40% focal lesions/none) and not chronic patients	BECHTER et al. 1987, 1989a
	CSF/serum antibody index	High prevalence of increased index (53%) or focal lesions (42%) or both (21%) among antibody carriers	BECHTER et al. 1989b
Acute Major depression (unipolar) Paranoid psychosis Dysthymia Personality disorder	Serum antibodies (follow-up samples)	Highest prevalence of antibody carriers (30%) among patients with major depression; sero-conversion in 50% of cases observed	BODE et al. 1992b, 1993a, b
Chronic Schizophrenia Paranoid psychosis Organic psychosyndrome Major depression (unipolar)	Serum antibodies (follow-up samples)	Lowest prevalence of antibody carriers (10%) among chronic schizophrenics	Bode et al. (unpublished data), also see this volume
	Antigen in peripheral blood monocytes (follow-up samples)	Similar high prevalence of antigen carriers (40–50%) among acute and chronic patients	BODE et al. 1994, also see this volume

Schizophrenia (two retrospective cases)	Serum antibodies	Coincident onset of disease with encephalitis (13 years ago, case 1) or major depression of a familymember (7 years ago, case 2), each of them now seropositive	BECHTER et al. 1992b
Neurologic diseases			
Acute meningoencephalitis (two cases)	Serum and CSF antibodies	Inoculation of patients' CSF produced transient foci in tissue culture and seroconversion, but no disease in rabbits	ROTT et al. 1991
Acute and chronic meningoencephalitis (with depression)	Serum antibodies	High prevalence of antibody carriers 70%) among some (unclear cases (n=17)	BECHTER et al. 1992c
Chronic encephalitis with dizziness, severe headache and depression		One case with exposure to seropositive farm animals	BECHTER et al. 1992a

BDV, Borna disease virus; CSF, cerebrospinal fluid; MRI, magnetic resonance imaging.

15%–20% antibody carriers (Bode et al. 1993a) and a high rate of 40%–50% antigen carriers (Bode et al. 1993c), as compared with (a limited number of) nonreactive blood donors tested by flow cytometry (Steinbach, personal communication). Certain syndromes were highly associated with evidence of BDV infection. Among the acute cohort (Fig. 5), the highest prevalence of antibody carriers (30%) was found in patients with (unipolar) major depression (Fig. 5a) (Bode et al. 1993a), whereas among the chronic patients (Fig. 6), the lowest rate (10%) was found in schizophrenics (Fig. 6a). These differences were also significant if both BDV-specific parameters were compared; no significant differences between patients' subcohorts were obtained if only antigen response was compared (Figs. 5b, 6b). It is not known whether other limbic dysfunctions that occur in diseases like chronic fatigue syndrome (Komaroff and Buchwald 1991) may also be related to BDV infection. In a limited study of these patients we have found no serologic evidence of BDV (Bode et al. 1992c); however, to be certain, further investigations, especially on BDV antigen response, are needed. The observation that viral antigen was found more frequently than antibodies to BDV and did not differ between acute and chronic patients, suggests that there may be subclinical infection in certain psychiatric syndromes with reactivation in acute disease episodes.

A variety of neurological disorders have been studied for evidence of BDV infection (Table 4). Bechter et al. (1992a, c) investigated single neurologic cases of unclear etiology and/or unusual course. Among 17 patients presenting with acute and chronic meningoencephalitis associated with dizziness, severe headache and depression, the majority (12%–70%) were seropositive. From two cases with acute meningoencephalitis, the inoculation of CSFs led to transient antigen foci in tissue culture and seroconversion without disease in rabbits (Rott et al. 1991). Although these results are not yet conclusive, they do, however, show that severe neurologic diseases should also be taken under consideration as possibly BDV-associated, if other causes are excluded.

Recently, Goldfarb and Gajdusek (1992) have reported on an unique, progressive, fatal neurologic disorder with unknown etiology which has occurred since 100 years among the Iakut people of Siberia and is named Viliuisk encephalomyelitis (VE). Many clinical features of VE amazingly resemble BD in horses and also the phasic course of BD in rats. The acute phase of VE, with fever, headache and lethargy, can lead to death in some cases, but more often is followed by a chronic phase with ataxia, stiffness and dementia, becoming fatal after 2–6 years. Behavioral and emotional changes are also commonly observed. In the brain of VE victims, inflammatory lesions, necrotic foci in the gray matter, and hydrocephalus could be detected. Although VE has never been suggested to be associated with BDV infection, it should be regarded as the "candidate of first choice" among particular neurologic diseases which are worthwhile to intensely investigate for BDV parameters in the future. As in natural BD, the mode of transmission in VE is still unknown; however, horizontal transmission within the population, together with an incubation time of up to several years, is suggestive (Goldfarb and Gajdusek 1992).

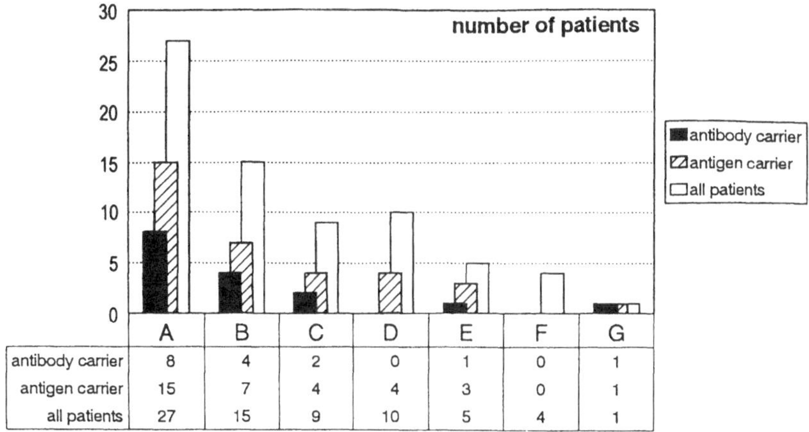

	A	B	C	D	E	F	G
antibody carrier	8	4	2	0	1	0	1
antigen carrier	15	7	4	4	3	0	1
all patients	27	15	9	10	5	4	1

A=unipolar depression, B=paranoid psychosis, C=personality disorder, D=neurotic/reactive depression (dysthymia) [7/3], E=schizophrenia/schizoaffective psychosis [1/4], F=bipolar depression, G=generalized anxiety disorder; follow-up study (6-weeks-period/patient); test of antibodies in serum and antigen in PBMs (peripheral blood monocytic cells)

a

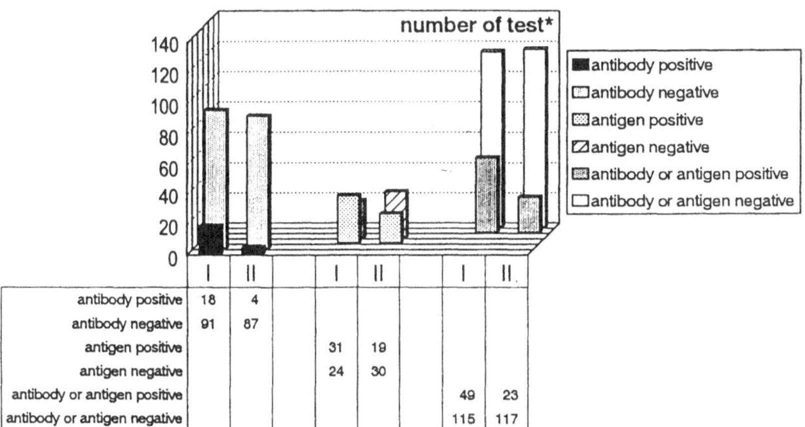

	I	II		I	II		I	II
antibody positive	18	4						
antibody negative	91	87						
antigen positive				31	19			
antigen negative				24	30			
antibody or antigen positive							49	23
antibody or antigen negative							115	117

I=unipolar depression, II=all other diagnoses except paranoid psychosis; *mean antibody tests/patient: I=4.0+-2.5, II=3.1+-2.5; mean antigen tests/patient: I=2.0+-1.5, II=1.7+-1.2; significant differences (chi-square-test) I versus II: antibody p<0.05, antigen non-sign., antibody and antigen p<0.01

b

Fig. 5a, b. Acute psychiatric patients with various syndromes (*A–G*) were analyzed for twofold BDV response (antibodies by immunofluorescence, antigen by fluorescence-activated cell sorting using blood samples collected during a follow-up of 6 weeks/patient, starting after onset of the acute disease episode.
Number of patients; **a** *A*, unipolar depression; *B*, paranoid psychosis; *C*, personality disorder; *D*, neurotic/reactive depression (dysthymia) (7/3); *E*, schizophrenia/schizoaffective psychosis (1/4); *F*, bipolar depression; *G*, generalized anxiety disorder; follow-up study (6 weeks/patient); antibodies in serum and antigen in peripheral blood monocytes tested. **b** Number of tests: I, unipolar depression (4.0 ± 2.5 antibody tests/patient, 2.0 ± 1.5 antigen tests/patient); II, all other diagnoses except paranoid psychosis (3.1 ± 2.5 antibody tests/patient, 1.7 ± 1.2 antigen tests/patient). Significant differences (χ^2 test) I vs II: antibody $p < 0.05$; antigen, not significant; antibody and antigen $p < 0.01$. Patients were hospitalized in the Dept. Psychiatry, Klinikum Steglitz, Free University, Berlin; clinical diagnoses were given by Prof. Dr. R. Ferszt, Dr. N. Rigas, Dr. H. Querfurth; the data on BDV antibody response were partially published (Bode et al. 1992b, 1993a,b)

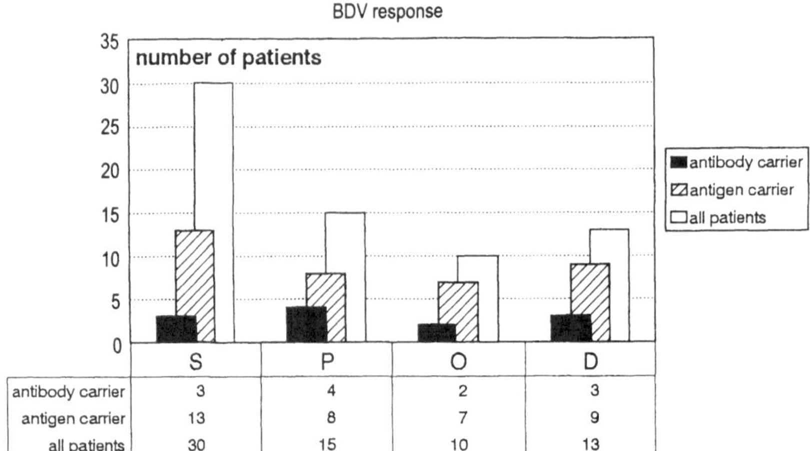

	S	P	O	D
antibody carrier	3	4	2	3
antigen carrier	13	8	7	9
all patients	30	15	10	13

S=schizophrenia, P=paranoid psychosis, O=organic psychosyndrome, D=depression (major depression [7] and schizoaffective psychosis [6] with depression as main symptom); follow-up study (one-year-period/patient); test of antibodies in serum and antigen in PBMs (peripheral blood monocytic cells)

a

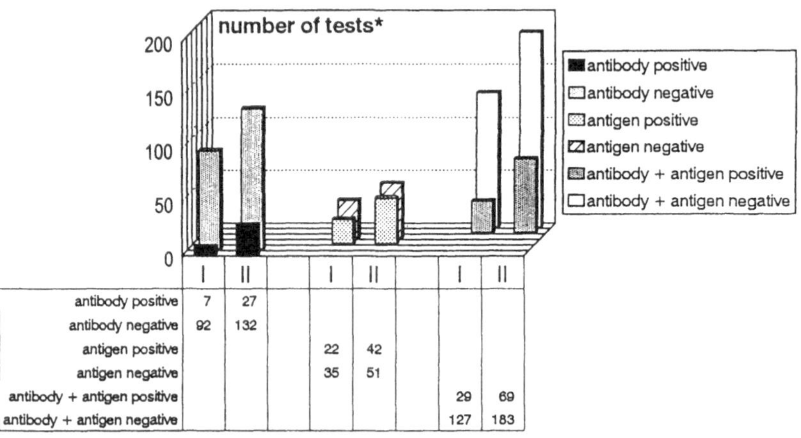

	I	II		I	II		I	II
antibody positive	7	27						
antibody negative	92	132						
antigen positive				22	42			
antigen negative				35	51			
antibody + antigen positive							29	69
antibody + antigen negative							127	183

I= schizophrenia, II= all other diagnoses;*mean antibody tests/patient: I=3.3+-1.6, II=4.2+-2.1; mean antigen tests/patient: I=1.9+-1.1, II=2.4+-1.2; significant differences (chi-square-test) I versus II: antibody p<0.025, antigen non-sign., antibody and antigen p<0.05

b

Fig. 6a, b. Chronic psychiatric patients with various syndromes (*S, P, O, D*) and a mean duration of illness of 20 ± 14 years were analyzed for twofold BDV response (antibodies by immunofluorescence, antigen by fluorescence-activated cell sorting) using blood samples collected during a follow-up of 1 year/patient.
Patients were hospitalized in the district hospital of Berlin-Zehlendorf; clinical diagnoses were given by Dr. G. Arkenberg, Dr. E.H. Kang-Welberts, Dr. Rommel.
a Number of patients; *S*, schizophrenia; *P*, paranoid psychosis; *O*, organic psychosyndrome; *D*, depression (seven with major depression and six with schizoaffective psychosis); antibodies in serum and antigen in peripheral blood monocytes tested. **b** Number of tests; I, schizophrenia (3.3 ± 1.6 antibody tests/patient, 1.9 ± 1.1 antigen tests/patient); II, all other diagnoses (4.2 ± 2.1 antibody tests/ patient; 2.4 ± 1.2 antigen tests/patient). Significant differences (χ^2 test) I vs II: antibody, $p < 0.025$; antigen not significant; antibody and antigen, $p < 0.05$.

5 Epidemiological Aspects

Intra-axonal transport has been shown to be the major route for dissemination of BDV in rabbits (KREY et al. 1979) and rats in which i.n. infection led to spread of the virus via the olfactory bulb to other brain areas (CARBONE et al. 1987; MORALES et al. 1988; GOSZTONYI et al. 1993). In natural infections of horses, the i.n. route appears to be the most likely portal of entry. Early reports had already given strong evidence that nasal secretions of BD horses can contain infectious virus (ZWICK et al. 1927). More recent attempts to isolate virus from such secretions of seropositive horses were negative (LANGE et al. 1987); however, even from the brain of BD horses it may be difficult to isolate infectious BDV (DÜRRWALD 1993).

The most important message from early and recent reports on natural animal hosts (Table 3) may be that inapparent infections probably occur often and that acute fatal BD might be a rather rare event. Indirect support for this conclusion was given by a comprehensive study on the prevalence of BD in horses and sheep in endemic areas of Eastern Germany (DÜRRWALD 1993). The maintenance of only sporadic BD cases among a large horse population leading to low mean prevalence rates of 0.1%–1% might be due to the existence of asymptomatic carriers which unrecognizably transmit the virus. Evidence for such carriers was shown by a previous report (LANGE et al. 1987), indicating 12% prevalence of immunoreactivity to BDV in horses without clinical signs in Western Germany.

For infection of human subjects, the modes of transmission remain an open question. Potential modes could include horizontal or vertical transmission within the human population, or transmission from infected animals. Exposure to seropositive farm animals was reported in one case of a BDV seropositive patient with acute encephalitis (BECHTER et al. 1992a), but the relationship remained speculative. Moreover, it seems rather unlikely that contact with farm animals can be a main source of human infections, since many infected subjects were detected among city populations.

Another possible reservoir for infection might be household pets. Antibodies to BDV were recently detected in Swedish cats suffering from an acute nonsuppurative meningoencephalitis called "staggering disease" (LUNDGREN and LUDWIG 1993; LUNDGREN et al. 1993). This cat disease has been known in Sweden for 20 years; the usually fatal syndrome is characterized by hind leg ataxia and paresis, fever, inability to retract the claws, and also depression and increased affection; inflammatory infiltrates are typically present in the gray matter of the brain stem (LUNDGREN 1992). Staggering disease shares many clinical features with BD in horses and sheep. Some 50% of animals with this disease (n=24) had BDV-specific antibodies (LUNDGREN and LUDWIG 1993). Studies of cats in Berlin (n=173) revealed that 7% had antibodies to BDV; among cats with undefined neurological disorders (n=24), 13% of animals had antibodies to BDV (LUNDGREN et al. 1993). Since the cat has become one of the most common household animals in developed countries and lives in closest contact to humans, its possible infection with BDV or a BDV-like agent would be of great clinical importance.

6 Conclusion

Borna disease virus is an unique, neurotropic, negative-stranded RNA virus which has a predilection for infecting limbic structures of the brain and causes various neurologic and behavioral syndromes in a broad spectrum of infected hosts. Asymptomatic carriers with life-long virus persistence are found in almost every susceptible species. Human infections appear to be likely, given specific serum antibodies against major BDV proteins (38/40 and 24kDa) encoded by the virus genome, and by the presence of BDV antigens in blood monocytes. The geographical distribution of human infections seems to be worldwide, since antibody carriers have been observed in Germany, North America, East Africa, and Japan. High seroprevalence rates (5%–15%) are exhibited among (hospitalized) patients with psychiatric, neurologic, and/or immunologic disorders, in contrast to low rates (1%–2%) among healthy volunteers. A role for BDV infection in human neuropsychiatric diseases is not yet conclusive. However, there are data to implicate BDV as important in affective disorders, since patients with major depression showed seroconversion and increased prevalence of both antibodies to BDV and of BDV antigen during acute disease episodes. The presence of antigen in PBMs indicates BDV infection irrespective of the host's actual antibody response; its increased frequency in certain psychiatric syndromes may be indicative of a relationship with disease. The increased prevalence of focal cerebral lesions in seropositive patients with acute major depression or schizophrenia suggests that BDV may cause encephalitis that presents as acute disease in psychiatric patients. Completely unknown is the mode of BDV transmission to humans. Potential modes could include such infected domestic animals as horses, sheep and cats. These animals need not have evidence of disease to serve as reservoir for BDV.

BDV is emerging as a potentially important human pathogen. We have yet to demonstrate viral genetic material in human cells, but this will now be considerably facilitated by our discovery of antigen-carrying cells in the blood. Apart from this, there is already a body of evidence to suggest that BDV could play a role in a variety of psychiatric and neurologic syndromes. The idea that viruses may be responsible for causing human brain diseases of unknown etiology is not new. A linkage has been proposed between influenza virus and schizophrenia since the pandemic in 1918 (MENNINGER 1928). More recently, CROW (1991) argued from genetic studies on families that schizophrenia may be due to an endogenous retrovirus. The data favoring BDV as a pathogen are more compelling. The fact that different neuropsychiatric diseases have been found to be associated with BDV suggests that the type of syndrome induced by infection may be a function of the host's particular hereditary vulnerability and the properties of the virus.

Acknowledgments. This work was supported by a research grant from the President of the Bundesgesundheitsamt (No. Fo 2-1322-665) given to the author and was partially supported by research grants from the Ministry of Research and Technology [BMFT] (No. 07017660), as well as from the Deutsche Forschungsgemeinschaft [DFG] (No. Lu 142/5-1) given to Prof. Dr. H. Ludwig, Institute of Virology, Free University (FU), 13353 Berlin, Germany. Special thanks are due to Prof. Dr. H. Ludwig,

Institute of Virology, (FU) Berlin, for providing the opportunity to work on this exiting aspect of the Bornavirus field, for lasting support and helpful discussions. I am grateful to P. Reckwald for excellent technical assistance. Furthermore, I like to thank F. Steinbach (Institute of Virology, FU Berlin) for providing the FACS pictures and some particular references, and G. Czech (from the same institute) for supporting the antibody studies. I also thank Dr. B. Thiele (Hospital of Internal Medicine, Charité, Humboldt University, Berlin) for offering free use of his FACScan. The following neurologists and psychiatrists essentially contributed to the success of this work with patients' blood specimens, descriptive clinical features, and helpful discussions: Dr. J.D. Amsterdam from the Depression Research Unit, Dept. Psychiatry, Univ. Pennsylvania, Philadelphia; Prof. Dr. R. Ferszt, Dr. H Berzewski, Dr. N. Rigas, and Dr. H. Querfurth from the Dept. Psychiatry, Klinikum Steglitz, FU Berlin; Dr. G. Arkenberg, Dr. E.H. Kang-Welberts, Dr. W Schwalbe, and Dr. Rommel from the District Hospital, Berlin-Zehlendorf; Dr. N. König and Dr. W. Pöllmann from the Marianne-Strauß-Klinik (Medical Center for MS) in Berg-Kempfenhausen; Dr. M. Haupts from the Dept. Neurology, Knappschaftskrankenhaus Bochum-Langendreer, University Hospital.

References

Amsterdam JD, Winokur A, Dyson, W, Herzog S, Gonzalez F, Rott R, Koprowski H (1985) Borna disease virus: a possible etiologic factor in human affective disorders? Arch Gen Psychiatry 42: 1093–1096

Amsterdam JD, Winokur A, Dyson W, Herzog S, Gonzalez F, Rott R, Koprowski H (1987) Demonstration of antibodies to Borna disease virus in patients with affective disorders. In: Kurstak E, Lipowski ZJ, Morozov PV (eds) Viruses, immunity and mental disorders. Plenum Medical, New York, pp 179–185

Amsterdam JD, Ludwig H, Riegel S, Bode L, Koprowski H (1988) Borna disease virus in psychiatric patients and healthy volunteers (abstract). 2nd world conference on viruses, immunity and mental health, 4–7 Oct 1988. Montreal, Canada

Anzil AP, Blinzinger K, Mayr A (1973) Persistent Borna virus infection in adult hamsters. Arch Gesamte Virusforsch 40: 52–57

Bause-Niedrig I, Pauli G, Ludwig H (1991) Borna disease virus-specific antigens: two different proteins identified by monoclonal antibodies. Vet Immunol Immunopathol 27: 293–301

Bechter K, Herzog S (1990) Über Beziehungen der Bornaschen Krankheit zu endogenen Psychosen. In: Kaschka WP, Aschauer HN (eds) Psychoimmunologie. Thieme, Stuttgart, pp 133–141

Bechter K, Herzog S, Fleischer B, Schüttler R, Rott R (1987) Kernspintomographische Befunde bei psychiatrischen Patienten mit und ohne Serum-Antikörper gegen das Virus der Bornaschen Krankheit. Nervenarzt 58: 617–624

Bechter K, Herzog S, Schüttler R, Rott R (1989a) MRI in psychiatric patients with serum antibodies against Borna disease virus. Psychiat Res 29: 281–282

Bechter K, Herzog S, Schüttler R, Rott R (1989b) Die Bornasche Krankheit-wahrscheinlich auch eine menschliche Krankheit: neue Ergebnisse. In: Saletu B (ed) Biologische Psychiatrie. Thieme, Stuttgart, pp 17–21

Bechter K, Schüttler R, Herzog S (1992a) Case of neurological and behavioral abnormalities: due to Borna disease virus encephalitis? Psychiatry Res 42: 193–196

Bechter K, Schüttler R, Herzog S (1992b) Borna disease virus: possible causal agent in psychiatric and neurological disorders in two families. Psychiatry Res 42: 291–294

Bechter K, Herzog S, Schüttler R (1992c) Possible significance of Borna diesease for humans. Neurol Psychiatr Brain Res 1: 23–29

Bode L, Ludwig H (1989) Borna disease virus infections and immune response in primates and man. In: Proceedings of the 3rd annual symposium of the European Society of Veterinary Neurology, Bern, Switzerland, pp 89–90

Bode L, Riegel S (1988) Immune response to Borna disease (BD) virus infections in animals and man (abstract). 2nd world conference on viruses, immunity and mental health, 4–7 Oct 1988; Montreal, Canada

Bode L, Riegel S, Ludwig H, Amsterdam JD, Lange W, Koprowski H (1988) Borna disease virus-specific antibodies in patients with HIV infection and with mental disorders. Lancet ii: 689

Bode L, Dittrich W, Ludwig H (1989) Lernschwäche bei klinisch gesunden Ratten nach Bornavirus-Infektion. Zentralbl Bakteriol 271: 394

Bode L, Riegel S, Czech G, Lange W, Ludwig H (1990a) Increased incidence of Borna disease virus infection in chronically diseased patients. In: VIIIth international congress of virology, abstracts, Berlin, Germany, p 138

Bode L, Riegel S, Reckwald P, Ludwig H (1990b) Improved and rapid serodiagnosis of Borna disease virus infections in animals and man. In : VIIIth international congress of virology, abstracts, Berlin, Germany, p 302

Bode L, Querfurth H, Ferszt R, Gosztonyi G, Rigas N, Czech G, Ludwig H (1991) Antibodies to Borna disease virus are four times as frequent in depressives as in controls. In: Wissenschaftswoche 1991, Forschungsprojekte am Klinikum Steglitz, Freie Universität Berlin, pp 269–271

Bode L, Riegel S, Lange W, Ludwig H (1992a) Human infections with Borna disease virus: seroprevalence in patients with chronic diseases and healthy individuals. J Med Virol 36: 309–315

Bode L, Czech G, Ferszt R, Ludwig H (1992b) Bornavirus-Infektion beim Menschen: eine neue Zoonose? In: Böhm R (ed) Bericht des 4. Hohenheimer Seminars "Aktuelle Zoonosen". Deutsche Veterinärmedizinische Gesellschaft, Schriftenreihe, Giessen (ISBN 3–924851–68–9), pp 138–147

Bode L, Komaroff AL, Ludwig H (1992c) No serologic evidence of Borna disease virus in patients with chronic fatigue syndrome. Clin Infect Dis 15: 1049

Bode L, Ferszt R, Czech G (1993a) Borna disease virus infection and affective disorders in man. Arch Virol Suppl 7: 159–167

Bode L, Ferszt R, Ludwig H (1993b) Ist ein "neues" Virus an affektiven Psychosen beteiligt? Hohe Prävalenz von Borna disease Virus (BDV) Antikörpern in psychiatrischen Patienten. In: 2. Deutscher Kongress für Infektions- und Tropenmedizin, abstracts, Berlin, Germany, p 43

Bode L, Steinbach F, Ludwig H (1994) A novel marker for Borna disease virus infection. Lancet 343: 297–298

Böyum A (1968) Isolation of mononuclear cells and granulocytes from human blood. Scand J Clin Lab Invest 21 [Suppl 97]: 77–89

Briese T, de la Torre JC, Lewis A, Ludwig H, Lipkin WI (1992) Borna disease virus, a negative-strand RNA virus, transcribes in the nucleus of infected cells. Proc Natl Acad Sci USA 89: 11486–11489

Carbone KM, Duchala CS, Griffin JW, Kincaid AL, Narayan O (1987) Pathogenesis of Borna disease in rats: evidence that intra-axonal spread is the major route for virus dissemination and determinant for disease incubation. J Virol 61: 3431–3440

Carbone KM, Moench T, Lipkin WI (1991) Borna disease virus replicates in astrocytes, Schwann cells and ependymal cells in persistently infected rats: location of viral genomic and messenger RNAs by in situ hybridization. J Neuropathol Exp Neurol 50: 205–214

Carter NP (1990) Measurement of cellular subsets using antibodies. In: Ormerod MG (ed) Flow cytometry, a practical approach. IRL Press at Oxford University Press, Oxford, pp 45–75

Cervos-Navarro J, Roggendorf W, Ludwig H, Stitz L (1981) Die Borna-Krankheit beim Affen unter Berücksichtigung der encephalitischen Reaktion. Verh Dtsch Ges Pathol 65: 208–212

Crow TJ (1991) The virogene hypothesis of psychosis: current status. In: Kurstak E (ed) Psychiatry and biological factors. Plenum, New York, pp 9–22

De la Torre JC, Carbone KM, Lipkin WI (1990) Molecular characterization of the Borna Disease agent. Virology 179: 853–856

Dittrich W, Bode L, Ludwig H, Kao M, Schneider K (1989) Learning deficiencies in Borna disease virus-infected but clinically healthy rats. Biol Psychiatry 26: 818–828

Dürrwald R (1993) Die natürliche Borna-Virus-Infektion der Einhufer und Schafe: Untersuchungen zur Epidemiologie, zu neueren diagnostischen Methoden (ELISA, PCR) und zur Antikörperkinetik bei Pferden nach Vakzination mit Lebendimpfstoff. Inagu dissertation, Fachbereich Veterinärmedizin, Freie Universität Berlin, 156 pages

Fu ZF, Amsterdam JD, Kao M, Shankar V, Koprowski H, Dietzschold B (1993) Detection of Borna disease virus-reactive antibodies from patients with affective disorders by Western immunoblot technique. J Affect Disord 27: 61–68

Goldfarb LG, Gajdusek DC (1992) Viliuisk encephalomyelitis in the Iakut people of Siberia. Brain 115: 961–978

Gosztonyi G, Ludwig H (1984a) Neurotransmitter receptors and viral neurotropism. Neuropsychiatr Clin 3: 107–114

Gosztonyi G, Ludwig H (1984b) Borna disease of horses: an immunohistological and virological study of naturally infected animals. Acta Neuropathol (Berl) 64: 213–221

Gosztonyi G, Briese T, Bode L, Lipkin WI, Ludwig H (1991) Immunohistological and molecular detection of Borna disease virus-specific structures in the brain. Clin Neuropathol 10: 262

Gosztonyi G, Dietzschold B, Kao M, Rupprecht CE, Ludwig H, Koprowski H (1993) Rabies and Borna disease: a comparative pathogenetic study of two neurovirulent agents. Lab Invest 68: 285–295

Haas B, Becht H, Rott R (1986) Purification and properties of an intranuclear virus-specific antigen from tissue infected with Borna disease virus. J Gen Virol 67: 235–241

Hirano N, Kao M, Ludwig H (1983) Persistent, tolerant or subacute infection in Borna disease virus infected rats. J Gen Virol 64: 1521–1530

Ihlenburg H, Brehmer H (1964) Beitrag zur latenten Borna-Erkrankung des Pferdes. Monatsh Vet Med 19: 463–465

Kao M, Gosztonyi G, Ludwig H (1983) Obesity syndrome in Borna disease virus infected rats. Zentralbl Bakteriol Mikrobiol Hyg [A] 255: 173

Kao M, Ludwig H, Gosztonyi G (1984) Adaptation of Borna disease virus to the mouse. J Gen Virol 65: 1845–1849

Kao M, Hamir AN, Rupprecht CE, Fu ZF, Shankar V, Koprowski H, Dietzschold B (1993) Detection of antibodies against Borna disease virus in sera and cerebrospinal fluid of horses in the USA. Vet Rec 132: 241–244

Komaroff AL, Buchwald D (1991) Symptoms and signs of chronic fatigue syndrome. Rev Infect Dis 13 [Suppl 1]: 8–11

Krey HF, Ludwig H, Rott R (1979) Spread of infectious virus along the optic nerve into the retina in Borna disease virus-infected rabbits. Arch Virol 61: 283–288

Lange H, Herzog S, Herbst W, Schliesser T (1987) Seroepidemiologische Untersuchungen zur Bornaschen Krankheit der Pferde. Tierarztl Umschau 42: 938–946

Lipkin WI, Carbone KM, Wilson MC, Duchala CS, Narayan O, Oldstone MBA (988) Neurotransmitter abnormalities in Borna disease. Brain Res 475: 366–370

Lipkin WI, Travis GH, Carbone KM, Wilson MC (1990) Isolation and characterization of Borna disease agent cDNA clones. Proc Natl Acad Sci USA 87: 4184–4188

Lipkin WI, Briese T, De la Torre JC (1992) Borna disease virus: molcular analysis of a neurotropic infectious agent (Mini-review). Microb Pathog 13: 167–170

Ludwig H, Becht H (1977) Borna disease — a summary of our present knowledge. In: ter Meulen V, Katz M (eds) Slow virus infections of the central nervous system, Springer, New York, Heidelberg, Berlin, pp 75–83

Ludwig H, Thein P (1977) Demonstration of specific antibodies in the central nervous system of horses naturally infected with Borna disease virus. Med Microbiol Immunol 163: 215–226

Ludwig H, Koester V, Pauli G, Rott R (1977) The cerebrospinal fluid of rabbits infected with Borna disease virus. Arch Virol 55: 209–223

Ludwig H, Kraft W, Kao M, Gosztonyi G, Dahme E, Krey HF (1985) Die Borna-Krankheit bei natürlich und experimentell infizierten Tieren: ihre Bedeutung für Forschung und Praxis. Tierarztl Prax 13: 421–453

Ludwig H, Bode L, Gosztonyi G (1988) Borna disease: a persistent virus infection of the central nervous system. Prog Med Virol 35: 107–151

Ludwig H, Furuya K, Bode L, Klein N, Dürrwald R, Lee DS (1993) Biology and neurobiology of Borna disease viruses (BDV), defined by antibodies, neutralizability and their pathogenic potential. Arch Virol Suppl 7: 111–133

Lundgren A-L (1992) Feline non-suppurative meningoencephalomyelitis. A clinical and pathological study. J Comp Pathol 107: 411–425

Lundgren A-L, Ludwig H (1993) Clinically diseased cats with non-suppurative meningoencephalomyelitis have Borna disease virus-specific antibodies. Acta Vet Scand 34: 101–103

Lundgren A-L, Czech G, Bode L, Ludwig H (1993) Natural Borna disease in domestic animals others than horses and sheep. J Vet Med B40: 298–303

Matthias D (1954) Der Nachweis von latent infizierten Pferden, Schafen und Rindern und deren Bedeutung als Virusreservoir bei der Borna'schen Krankheit. Arch Exp Vet Med 8: 506–511

Menninger KA (1928) The schizophrenia syndrome as the product of infectious disease. Arch Neurol Psychiatry 20: 464–481

Morales JH, Herzog S, Kompter C, Frese K, Rott R (1988) Axonal transport of Borna disease virus along olfactory pathways in spontaneously and experimentally infected rats. Med Microbiol Immunol 177: 51–68

Narayan O, Herzog S, Frese K, Scheefers K, Rott R (1983a) Pathogenesis of Borna disease in rats: immune-mediated viral ophthalmoencephalopathy causing blindness and behavioral abnormalities. J Infect Dis 148: 305–315

Narayan O, Herzog S, Frese K, Scheefers H, Rott R (1983b) Behavioral disease in rats caused by immunopathological responses to persistent Borna virus in the brain. Science 220: 1401–1403

Nicolau S, Galloway IA (1928) Borna disease and enzootic encephalomyelitis of sheep and cattle. Spec Rep Med Res Council 121: 7–90

Pauli G, Gregersen J-P, Ludwig H (1984) Plaque/focus-immunoassay: a simple method for detecting antiviral or other antibodies and viral antigens in cells. J Immunol Methods 74: 337–344

Pette H, Környey S (1935) Über die Pathogenese und die Pathologie der Borna'schen Krankheit im Tierexperiment. Dtsch Z Nervenheilkd 136: 20–65

Pyper JM, Richt JA, Brown L, Rott R, Narayan O, Clements JE (1993) Genomic organization of the structural proteins of Borna disease virus revealed by a cDNA clone encoding the 38 kDa protein. Virology 195: 229–238
Richt JA, Stitz L (1992) Borna disease virus infected astrocytes function in vitro as antigen-presenting and target cells for virus-specific CD4-bearing lymphocytes. Arch Virol 124: 95–109
Richt JA, Stitz L, Wekerle H, Rott R (1989) Borna disease, a progressive meningoencephalomyelitis as a model for CD4+ T cell-mediated immunopathology in the brain. J Exp Med 170: 1045–1050
Richt JA, Stitz L, Deschl U, Frese K, Rott R (1990) Borna disease virus-induced meningoencephalomyelitis caused by a virus-specific CD4+ T cell-mediated immune reaction. J Gen Virol 71: 2565–2573
Richt JA, Vande Woude S, Zink MC, Narayan O, Clements JE (1991) Analysis of Borna disease virus-specific RNAs in infected cells and tissues. J Gen Virol 72: 2252–2255
Rott R, Herzog S, Fleischer B, Winokur H, Amsterdam JD, Dyson W, Koprowski H (1985) Detection of serum antibodies to Borna disease virus in patients with psychiatric disorders. Science 228: 755–756
Rott R, Herzog S, Bechter K, Frese K (1991) Borna disease, a possible hazard for man? Arch Virol 118: 143–149
Seifried O, Spatz H (1930) Die Ausbreitung der encephalitischen Reaktion bei der Borna'sche Krankheit der Pferde und deren Beziehungen zu der Encephalitis epidemica, der Heine-Medinschen Krankheit und der Lyssa des Menschen. Eine vergleichend-pathologische Studie. Zentralbl Neurol Psychiatry 124: 317–382
Shankar V, Kao M, Hamir AN, Sheng H, Koprowski H, Dietzschold B (1992) Kinetics of virus spread and changes in levels of several cytokine mRNAs in the brain after intranasal infection of rats with Borna disease virus. J Virol 66: 992–998
Sierra-Honigmann AM, Rubin SA, Estafanous MG, Yolken RH, Carbone KM (1993) Borna disease virus in peripheral blood mononuclear and bone marrow cells of neonatally and chronically infected rats. J Neuroimmunol 45: 31–36
Sprankel H, Richarz K, Ludwig H, Rott R (1978) Behavioral alterations in tree shrews (Tupaia glis, Diard 1820) induced by Borna disease virus. Med Microbiol Immunol 165: 1–18
Steinbach F, Thiele B Bode L, Zimmermann W, Ludwig H (1993) Borna disease virus is carried in monocytic cells during infection of animals (submitted)
Steinbach F, Thiele B, (1993) Phenotypical investigation of mononuclear phagocytes by flow cytometry. J Immunol Methods 174: 109–122.
Stitz L, Krey HF, Ludwig H (1980) Borna disease in rhesus monkeys as a model for uveo-cerebral symptoms. J Med Virol 6: 333–340
Stitz L, Soeder D, Deschl U, Frese K, Rott R (1989) Inhibition of immune-mediated meningoencephalitis in persistently Borna disease virus infected rats by cyclosporine A. J Immunol 143: 4250–4256
Stitz L, Schilken D, Frese K (1991) Atypical dissemination of the highly neurotropic Borna disease virus during persistent infection in cyclosporin A-treated, immunosuppressed rats. J Virol 65: 457–460
Vande Woude S, Richt JA, Zink MC, Rott R, Narayan O, Clements JE (1990) A Borna virus cDNA encoding a protein recognized by antibodies in humans with behavioral diseases. Science 250:1278–1281
Zwick W (1939) Bornasche Krankheit und Encephalomyelitis der Tiere. In: Gildemeister F, Hagen E, Waldmann O (eds) Handbuch der Viruskrankheiten, vol 2. Fischer, Jena, pp 254–354
Zwick W, Seifried O, Witte J (1927) Experimentelle Untersuchungen über die seuchenhafte Gehirn-Rückenmarkentzündung der Pferde (Borna'sche Krankheit). Z Inf Krkh Haustiere 30: 42–136

Note Added in Proof

During the publishing process of this book (acceptance of this article in November 1993, proofs in October 1994), we succeeded to demonstrate BDV genomic transcripts together with antigen in peripheral blood leucocytes from psychiatric patients and sequence data of a part of the viral gene encoding the 38/40 kDa protein. This is the first definite proof that a human Borna disease virus exists (Bode L, Zimmermann W, Ferszt R, Steinbach F, Ludwig H (1995) Borna disease virus genome transcribed and expressed in psychiatric patients. Nature Medicine 1 (3): 232–236).

Subject Index

acute disease, psychiatric 121
- focal cerebral lesions 121
- follow-up sampling 121
acute/subacute infection 60
adoptive transfer 80
adventitial infiltrates 60
affective disorders 121, 126
aggressivity 94
amygdalal area 97, 98
animal model 24
antagonist, D1 receptor 96
antibodies 8, 20–22, 104, 109
- assay of 22
- detection of 109
- monoclonal 104, 109
- neutralizing 20, 109
- polyclonal 104
antibody carriers 120
antigen-carrying PBMs 116
antigens 52, 53
- in peripheral blood monocytes (see PBMs) 112, 113, 116
- phasic expression 53
- s-(soluble) 108
- three types 52
- virus-specific (*see also there*) 48, 63
aspartate 63, 64
astrocytes 44, 57, 59, 70, 82
asymptomatic carriers 120, 125
attention deficit disorders (ADD) 99

B cells 78
BDV antigens (*see* antigens)
BDV-specific monoclonal antibody (*see also there*) 109
behavioral disorders / abnormalities 24, 25, 79, 93 ff., 121
- biphasic course 121
- ingestive 98
- mating behaviors 94
- motor behaviors 93
- observations 93
- rats 121
- sexual 98
- social behaviors 93, 94
- stereotyped 95
- tree shrews 121
Borna disease in Saxony 77
Borna-like disease of Ostriches in Israel 31 ff.
brain diseases of horses 93

carriers of virus 22
- asymptomatic 125
cats 46
cattle 23
caudate-putamen 96
CD4+ 82, 87
- helper/inflammatory 82
- T cells 84, 86
CD8+ 82, 84, 88
- cytotoxic 86
- pathogenetic relevance 84
- T cells 84, 85, 87
cDNAs 2, 3, 9
cell and tissue tropism of BDV 64
cellular immune response 67, 82
characterization 11
chicken 23, 62
cholinergic neurons 98
clinical symptoms 21, 23
clozapine 97
CNS 40, 75 ff.
- diseases in 75 ff.
- inflammatory reaction 40
Creutzfeld-Jakob-Scheinker syndrome 76
CSFs 111
- horses 112
- rabbits 112
CTL (cytotoxic lymphocytes) 86, 88
cyclophophamide 80
cyclosporine A (CSA) 80, 81
- immunosuppressed rats 81
cytokines / cytokine treatment 84–87
- immunopathogenesis 84
- inflammatory 86
- proinflammatory 87
cytopathogenicity 77, 81
- in vitro 77

cytotoxic lymphocytes (*see* CTL)
cytotoxicity / cytotoxic effects 81, 83, 86

d-amphetamine 95
D1 receptor antagonist 96
debility 86
degeneration, virus-induced 65
delayed-type hypersensitivity reaction (DTH) 83, 84
dementia 86
dendritic cells 84
dentate gyrus 50, 60, 62, 66
depression, major 122
diencephalon 46
disease of the head 77
dissemination virus, extraneural 80
DNA
– cDNAs (*see there*) 2, 3, 9
dopamine
– agonist 95
– disturbances 96
– metabolite 96
– neurons 95
– system 95
– – prefrontal 96
double stain technique, human sera 109
dyskinesia, tardive 97
dystonia 95

electron microscopy 45, 58
– horse brain with BD 45
ELISA 33, 34
encephalitis/encephalitic lesions 41, 45, 46, 48, 78–80, 84
– horse 41
– nonpurulent 46
encephalomyelitis 103, 123
– nonpurulent 103
– – horses 103
– – sheep 103
– Viliuisk 123
endemic areas 103
enzyme immunoassays 109
epidemiological aspects 125
experimental infections/transmission 23, 47
extraneural organs 65
extrapyramidal syndrome 94

FACS (fluorescence-activated cell sorting) 112
fatal neurologic disorder 123
flow cytometric detection 113
frontal lobe syndrome 94, 95

glutamate 63, 64
grooming 95
guinea pigs 23

hamsters 62
hippocampal
– area 98
– formation 50, 63
– pathology 95
hippocampus 46, 50, 51, 66, 97
histopathology 21, 24, 25, 47
horse brain 43
horses 93, 103, 112
– blepharospasm 93
– trismus 93
– twitching 93
hospitalized patients 126
– immunologic disease 126
– neurologic disease 126
– psychiatric disease 126
host range/species 19, 20, 23, 24, 26, 120
– broad 104
– different 120
human / humoral response 99, 108, 109, 111, 126
– diseases 99
– pathogen 126
– sera 109, 111
hydrocephalus 85
hyperactivity 93–95, 99
hyperacute infection 59
hyperphagia 66
hypersensitivity reaction, delayed-type (DTH) 83, 84
hypothalamus 52, 62, 66

i. n. route 125
IFN (interferon) 83, 85
– IFN-α/β 85
– IFN-β 88
– IFN-γ 83, 85, 88
IL (interleukin)
– IL-1 87, 88
– IL-2 87
– IL-6 87
immunocompromised rats 80
immunoelectron microscopy (*see also* electron microscopy) 49, 59, 63
immunofluorescence assays 33, 37, 108
– human sera 109
– MA104 cells 33
immunologic disease, hospitalized patients 126
immunopathogenesis 75ff., 81, 82
– cytokine treatment 84
immunoprecipitation 109
immunosuppression 80, 81
– pharmacological 80
in situ hybridization 44, 55, 62, 98
in vivo marker 115
inapparent infections 125
inclusion bodies 42
inducible nitric oxide synthase (iNOS) 87
infection/BDV infection
– intracerebral 110
– intranasal 110
– subclinical 22

inflammatory
- cells 82
- changes 58
- reactions 78
infundibulum 58, 66, 67
- infundibular region 66, 67
ingestive behavior 98
inoculations
- intracerebral 60
- intraocular 60
- peripheral 60
iNOS (inducible nitric oxide synthase) 87
insect transmission 22
interferon (*see* IFN)
interleukin (*see* IL)
intra-axonal transport 125

Joest-Degen inclusion bodies 44, 45, 49

Klüver-Bucy syndrome 66, 67
Kuru-Kuru 76

learning 79, 95
- deficiency 65
leptomeningeal infiltrates 60
Lesch-Nyhan patients 96
leukocytes, polymorphonuclear 40
limbic
- cortical areas 98
- structures 96, 97, 104
- system 78
locomotor activation 96, 97

macrophage activation 58
macrophages 78, 84, 87
mad-cow disease 76
major
- BDV proteins 108
- depression 122
mating behaviors 94
meningoencephalitis 123
mesencephalon 46
mesocortical system 95
- dopamine system 96
mesolimbic system 95
- dopamine system 96
mesostriatal system 95
MHC antigens 82 ff.
- class I 82, 85, 86
- class II 82, 83, 85
MHC-expression cells 82
mice 62, 94
microglia 41
molecular biology of Borna disease virus 1 ff.
monkey 62, 104
monoclonal antibodies 104, 109
- BDV-specific 109
monocytes 87
mortality 21
movement disorders 93
mRNA 3, 9

natural form 40
negative-strand
- genomic RNA 104
- virus 5
neuroanatomy 93, 97
neurobehavioral syndrome, subtle 103
neurochemistry 93, 95
neurologic
- disease 97, 120, 123, 126
- - hospitalized patients 126
- - various 120
- disorder, fatal 123
neuronal
- damage 42, 85
- degeneration 55
- destruction 86
neurons 78, 82, 95
- dopamine 95
neuropathological picture of syndrome 61
neurophagic nodules 42
neuropharmacology 93
neuropsychiatric diseases 93
- associated with BDV 127
neuropsychological deficits 55
neurotoxins 85, 86
neurotransmitter
- abnormalities 121
- systems 63
neurotropic agent 65, 104
neutralizing antibodies 109
nigrostriatal dopamine system 96
noncytopathic virus 77
nuclear
- fluorescence, human sera 109
- phase 6
nucleic acids 43
nucleus
- accumbens 96
- paraventricular 62, 66, 67

obesity
- syndrome, BDV induced 61, 66
- virus-induced 67
oligodendrocytes 44, 53, 57, 78
ostriches 46
- in Israel, Borna-like disease of 31 ff.

panencephalitis 46
paraparesis of hind limbs 47
paraventricular nucleus 62, 66, 67
parenchymal damage, virus-induced 55
paresis, clinical signs 35
pathoclisis 65
pathogenesis 20, 26, 63
- immunopathology 26
- virus entry 20
- virus spread 20
PBMs (peripheral blood monocytes 112, 113, 116
- antigen, horses 112, 113

PBMs (peripheral blood monocytes
– antigen-carrying 116
persistence/persistent infection 18, 19, 22, 103
– in brain cells 18, 22
– in cell culture 19
pharmacological immunosuppressants 80
pharmacology 95
phylogenetically distant 77
polioencephalitides, nonpurulent 45
polymorphonuclar leukocytes 40
portal of entry of BDA 63
protein
– 14,5 kDa 9, 11
– 24 kDa, p24 3, 6, 8–10
– 38/40 kDa, p40 3, 6, 8–10
– 60 kDa 8
– major 108
– nuclear localization signal 10
– phosphorylation 8
– recombinant p24 35
– recombinant p 38 35
– s-antigen 8, 10
psychiatric diseases 113, 121, 126
– acute 113
– hospitalized patients 126
– psychiatric out-patients 120
psychosis 99
Purkinje cells 52

rabbit 62, 112
rabbit-adapted virus 47
rabies virus 63
radicals 85
rat 47, 77, 79, 80, 93–95, 104, 121
– athymic 79
– immunocompromised 80
– newborn 79
reactive changes 58
receptors 63
reservoir for BDV 126
retina 42, 54, 57
retinitis 80
retrocollis 95
rhesus monkey 80, 104
RNA
– genomic 5, 7
– mRNA (*see there*) 3, 9
– negative-stranded RNA 40, 104
– polarity 3, 5
– polyadenylation, poly A +/– 3, 7
– single-stranded RNA virus 3, 40
– transcripts 3, 8

schizophrenia 99, 123
Schwann cells 55, 59, 78
Scrapie agent 76
self-mutilation 95, 96
sera, human 109, 111
serological methods 108
– immunofluorescence assays 108
sexual behavior 98
sheep 23, 46, 94, 103
shrews 62, 94
single-stranded RNA virus 3
social behavior 93, 94
social interactions 94
soluble (s-) antigen 50
spectrum of different diseases 116
spinal cord 54
spread of BD (*see* virus-spread)
staggering disease 126
stereotypy 95
striatal areas 96
subacute infection 60
subclinical infections 22
subtle neurobehavioral syndromes 103
symptoms
– behavioral disorders (*see also there*) 24, 25, 93 ff.
– clinical 21, 23

T cells 78–83
TGF-treated BDV 85
thalamus 52, 98
tissue tropism of BDV 64
TNF (tumor necrosis factor) 87, 88
torticollis 94
Tourette's syndrome 99
transmission
– experimental 23
– by insects 22
– by secretions 19, 22
tree shrews 94, 104, 121
tumor necrosis factor (*see* TNF) 87
Tupaia glis 94, 96

vacuolar myelopathy 57
variability 25
various neurologic disease 120
Viliuisk encephalomyelitis 123
viral dissemination in peripheral organs 115
viral proteins, phasic expression of 66
virus-induced
– degeneration 65
– parenchymal damage 55
virus-specific
– antigen 43, 48, 63
– – distribution 48
– – virus-spread 48
– immune response 76
virus-spread of BD 50, 54, 63
– axonal 50, 63
– centrifugally 54
– natural infections 63
– transsynaptic (transneuronal) 50, 63
visceral organs 55
vulnerability, elective 65

western blot 109

Current Topics in Microbiology and Immunology

Volumes published since 1989 (and still available)

Vol. 147: **Vogt, Peter K. (Ed.):** Oncogenes. Selected Reviews. 1989. 8 figs. VII, 172 pp. ISBN 3-540-51050-8

Vol. 148: **Vogt, Peter K. (Ed.):** Oncogenes and Retroviruses. Selected Reviews. 1989. XII, 134 pp. ISBN 3-540-51051-6

Vol. 149: **Shen-Ong, Grace L. C.; Potter, Michael; Copeland, Neal G. (Ed.):** Mechanisms in Myeloid Tumorigenesis. Workshop at the National Cancer Institute, National Institutes of Health, Bethesda, MD, USA, March 22, 1988. 1989. 42 figs. X, 172 pp. ISBN 3-540-50968-2

Vol. 150: **Jann, Klaus; Jann, Barbara (Ed.):** Bacterial Capsules. 1989. 33 figs. XII, 176 pp. ISBN 3-540-51049-4

Vol. 151: **Jann, Klaus; Jann, Barbara (Ed.):** Bacterial Adhesins. 1990. 23 figs. XII, 192 pp. ISBN 3-540-51052-4

Vol. 152: **Bosma, Melvin J.; Phillips, Robert A.; Schuler, Walter (Ed.):** The Scid Mouse. Characterization and Potential Uses. EMBO Workshop held at the Basel Institute for Immunology, Basel, Switzerland, February 20–22, 1989. 1989. 72 figs. XII, 263 pp. ISBN 3-540-51512-7

Vol. 153: **Lambris, John D. (Ed.):** The Third Component of Complement. Chemistry and Biology. 1989. 38 figs. X, 251 pp. ISBN 3-540-51513-5

Vol. 154: **McDougall, James K. (Ed.):** Cytomegaloviruses. 1990. 58 figs. IX, 286 pp. ISBN 3-540-51514-3

Vol. 155: **Kaufmann, Stefan H. E. (Ed.):** T-Cell Paradigms in Parasitic and Bacterial Infections. 1990. 24 figs. IX, 162 pp. ISBN 3-540-51515-1

Vol. 156: **Dyrberg, Thomas (Ed.):** The Role of Viruses and the Immune System in Diabetes Mellitus. 1990. 15 figs. XI, 142 pp. ISBN 3-540-51918-1

Vol. 157: **Swanstrom, Roland; Vogt, Peter K. (Ed.):** Retroviruses. Strategies of Replication. 1990. 40 figs. XII, 260 pp. ISBN 3-540-51895-9

Vol. 158: **Muzyczka, Nicholas (Ed.):** Viral Expression Vectors. 1992. 20 figs. IX, 176 pp. ISBN 3-540-52431-2

Vol. 159: **Gray, David; Sprent, Jonathan (Ed.):** Immunological Memory. 1990. 38 figs. XII, 156 pp. ISBN 3-540-51921-1

Vol. 160: **Oldstone, Michael B. A.; Koprowski, Hilary (Eds.):** Retrovirus Infections of the Nervous System. 1990. 16 figs. XII, 176 pp. ISBN 3-540-51939-4

Vol. 161: **Racaniello, Vincent R. (Ed.):** Picornaviruses. 1990. 12 figs. X, 194 pp. ISBN 3-540-52429-0

Vol. 162: **Roy, Polly; Gorman, Barry M. (Eds.):** Bluetongue Viruses. 1990. 37 figs. X, 200 pp. ISBN 3-540-51922-X

Vol. 163: **Turner, Peter C.; Moyer, Richard W. (Eds.):** Poxviruses. 1990. 23 figs. X, 210 pp. ISBN 3-540-52430-4

Vol. 164: **Bækkeskov, Steinnun; Hansen, Bruno (Eds.):** Human Diabetes. 1990. 9 figs. X, 198 pp. ISBN 3-540-52652-8

Vol. 165: **Bothwell, Mark (Ed.):** Neuronal Growth Factors. 1991. 14 figs. IX, 173 pp. ISBN 3-540-52654-4

Vol. 166: **Potter, Michael; Melchers, Fritz (Eds.):** Mechanisms in B-Cell Neoplasia. 1990. 143 figs. XIX, 380 pp. ISBN 3-540-52886-5

Vol. 167: **Kaufmann, Stefan H. E. (Ed.):** Heat Shock Proteins and Immune Response. 1991. 18 figs. IX, 214 pp. ISBN 3-540-52857-1

Vol. 168: **Mason, William S.; Seeger, Christoph (Eds.):** Hepadnaviruses. Molecular Biology and Pathogenesis. 1991. 21 figs. X, 206 pp. ISBN 3-540-53060-6

Vol. 169: **Kolakofsky, Daniel (Ed.):** Bunyaviridae. 1991. 34 figs. X, 256 pp. ISBN 3-540-53061-4

Vol. 170: **Compans, Richard W. (Ed.):** Protein Traffic in Eukaryotic Cells. Selected Reviews. 1991. 14 figs. X, 186 pp. ISBN 3-540-53631-0

Vol. 171: **Kung, Hsing-Jien; Vogt, Peter K. (Eds.):** Retroviral Insertion and Oncogene Activation. 1991. 18 figs. X, 179 pp. ISBN 3-540-53857-7

Vol. 172: **Chesebro, Bruce W. (Ed.):** Transmissible Spongiform Encephalopathies. 1991. 48 figs. X, 288 pp. ISBN 3-540-53883-6

Vol. 173: **Pfeffer, Klaus; Heeg, Klaus; Wagner, Hermann; Riethmüller, Gert (Eds.):** Function and Specificity of γ/δ TCells. 1991. 41 figs. XII, 296 pp. ISBN 3-540-53781-3

Vol. 174: **Fleischer, Bernhard; Sjögren, Hans Olov (Eds.):** Superantigens. 1991. 13 figs. IX, 137 pp. ISBN 3-540-54205-1

Vol. 175: **Aktories, Klaus (Ed.):** ADP-Ribosylating Toxins. 1992. 23 figs. IX, 148 pp. ISBN 3-540-54598-0

Vol. 176: **Holland, John J. (Ed.):** Genetic Diversity of RNA Viruses. 1992. 34 figs. IX, 226 pp. ISBN 3-540-54652-9

Vol. 177: **Müller-Sieburg, Christa; Torok-Storb, Beverly; Visser, Jan; Storb, Rainer (Eds.):** Hematopoietic Stem Cells. 1992. 18 figs. XIII, 143 pp. ISBN 3-540-54531-X

Vol. 178: **Parker, Charles J. (Ed.):** Membrane Defenses Against Attack by Complement and Perforins. 1992. 26 figs. VIII, 188 pp. ISBN 3-540-54653-7

Vol. 179: **Rouse, Barry T. (Ed.):** Herpes Simplex Virus. 1992. 9 figs. X, 180 pp. ISBN 3-540-55066-6

Vol. 180: **Sansonetti, P. J. (Ed.):** Pathogenesis of Shigellosis. 1992. 15 figs. X, 143 pp. ISBN 3-540-55058-5

Vol. 181: **Russell, Stephen W.; Gordon, Siamon (Eds.):** Macrophage Biology and Activation. 1992. 42 figs. IX, 299 pp. ISBN 3-540-55293-6

Vol. 182: **Potter, Michael; Melchers, Fritz (Eds.):** Mechanisms in B-Cell Neoplasia. 1992. 188 figs. XX, 499 pp. ISBN 3-540-55658-3

Vol. 183: **Dimmock, Nigel J.:** Neutralization of Animal Viruses. 1993. 10 figs. VII, 149 pp. ISBN 3-540-56030-0

Vol. 184: **Dunon, Dominique; Mackay, Charles R.; Imhof, Beat A. (Eds.):** Adhesion in Leukocyte Homing and Differentiation. 1993. 37 figs. IX, 260 pp. ISBN 3-540-56756-9

Vol. 185: **Ramig, Robert F. (Ed.):** Rotaviruses. 1994. 37 figs. X, 380 pp. ISBN 3-540-56761-5

Vol. 186: **zur Hausen, Harald (Ed.):** Human Pathogenic Papillomaviruses. 1994. 37 figs. XIII, 274 pp. ISBN 3-540-57193-0

Vol. 187: **Rupprecht, Charles E.; Dietzschold, Bernhard; Koprowski, Hilary (Eds.):** Lyssaviruses. 1994. 50 figs. IX, 352 pp. ISBN 3-540-57194-9

Vol. 188: **Letvin, Norman L.; Desrosiers, Ronald C. (Eds.):** Simian Immunodeficiency Virus. 1994. 37 figs. X, 240 pp. ISBN 3-540-57274-0

Vol. 189: **Oldstone, Michael B. A. (Ed.):** Cytotoxic T-Lymphocytes in Human Viral and Malaria Infections. 1994. 37 figs. IX, 210 pp. ISBN 3-540-57259-7

Springer-Verlag and the Environment

We at Springer-Verlag firmly believe that an international science publisher has a special obligation to the environment, and our corporate policies consistently reflect this conviction.

We also expect our business partners – paper mills, printers, packaging manufacturers, etc. – to commit themselves to using environmentally friendly materials and production processes.

The paper in this book is made from low- or no-chlorine pulp and is acid free, in conformance with international standards for paper permanency.